Practical Research Methods for Nonprofit and Public Administrators

Practical Research Methods for Nonprofit and Public Administrators

ELIZABETHANN O'SULLIVAN

North Carolina State University

GARY R. RASSEL

University of North Carolina-Charlotte

JOCELYN DEVANCE TALIAFERRO

North Carolina State University

Routledge
Taylor & Francis Group

LONDON AND NEW YORK

First published 2011 by Pearson Education, Inc.

Published 2016 by Routledge
2 Park Square, Milton Park, Abingdon, Oxon OX14 4RN
711 Third Avenue, New York, NY, 10017, USA

Routledge is an imprint of the Taylor & Francis Group, an informa business

ISBN-13: 978-0-205-63946-5 (pbk)

Cover Design Manager: Jayne Conte
Cover Designer: Suzanne Duda

Library of Congress Cataloging-in-Publication Data
O'Sullivan, Elizabethann.
Practical research methods for nonprofit and public administrators / Elizabethann O'Sullivan,
Gary R. Rassel, Jocelyn DeVance Taliaferro.
p. cm.
Includes index.
ISBN-13: 978-0-205-63946-5
1. Human services—Research—Methodology. 2. Social service—Research—Methodology.
3. Nonprofit organizations—Research—Methodology. 4. Public administration—Research—
Methodology. I. Rassel, Gary Raymond. II. Taliaferro, Jocelyn DeVance. III. Title.
HV11.O79 2011
001.4'2—dc22   2010032090

TABLE OF CONTENTS

PREFACE

I magine how disheartened we were when students told us how much they learned in their research methods class, but they did not think that it would be relevant to their careers. We suppressed the urge to say, "But you will use it all the time." Students may not realize how relevant research methods will be in their careers, and administrators may not appreciate when they are applying research skills. If we are right, leaders and staff in public and nonprofit organizations are missing opportunities to use research information effectively and efficiently. We immediately saw the need to explicitly link management applications and research methodology, which led to this text.

In writing this book we included four innovations to demonstrate the link between methodology and professional practice. First, we integrate traditional research methods topics with specific management applications. Second, we include extensive exercises, which require students to apply each methodology. Third, we limit the methodological details to the key concepts that administrators and managers use most often. Fourth, we acknowledge the importance of qualitative methods and emphasize practical skills that managers should be able to easily and correctly apply.

The senior authors, Elizabethann O'Sullivan and Gary Rassel, have written together for over 25 years. This text includes a third author, Jocelyn Taliaferro. Jocelyn brings in the point of view of another discipline, social work, which complements our backgrounds in public and nonprofit management. More importantly, she is a trained qualitative analyst. Her voice in the book should encourage you to see the value of these methods, to use them correctly, and get the greatest value of the information they provide. Jocelyn's participation in writing this text makes it truly interdisciplinary and relevant to professionals who come from a variety of academic backgrounds.

Social science and policy researchers rely on measurement, sampling, and research designs to create and conduct their research. In addition, they must apply statistics and other analytical procedures to analyze and interpret the information they collect. All the management tools we selected require measurement, sampling, and a choice of a research design, but one or two methodologies dominate each tool. Consequently, we organized this text around management tools that rely heavily on methodology: performance measurement and monitoring, citizen and client surveys, program evaluation, and needs assessments. For each tool we also introduce the basic statistics used to organize and interpret the data.

The exercises are our central pedagogical tool. We have observed that students learn more from hands-on applications than they do from lectures. Plus, the exercises provide immediate feedback about what is understood and what is unclear. We divided the exercises into two or three parts: Getting Started, Small Group Exercises, and On Your Own. Getting Started asks students to apply the information presented in the chapter to management situations or policy problems. The Small Group or Class Exercises mirror what happens in most research projects, that is, each individual participant has ideas about what the research question is, why it was asked, and how to answer it. Exchanging and challenging ideas and resolving differences are critical to effective studies. The opportunity to practice these skills should be valuable in students' future careers. As for now, students will find that as they try to explain their ideas, listen to others' ideas, and incorporate different insights they will understand the methodology better. An On Your Own exercise is included in most chapters. This part of the exercises guides students who are working or have an internship through the steps required to apply the methodology in the field.

Our hardest decisions were how much detail to include. Our discussions of the management tools are limited—one can't begin to learn everything in one course. In the methodology chapters we did not cover the more technical details. We assumed that readers who plan careers as researchers or policy analysts will take advanced courses in methodology and statistics. We confined ourselves to covering the information needed to be an effective manager.

Including qualitative methods in a methodology text may not seem unique, but instructors may often skip the qualitative chapters as they and their students focus on the quantitative components designed to produce "objective," verifiable information. Quantitatively trained authors may present qualitative methods as an add-on to quantitative studies. These authors may unconsciously communicate their limited knowledge of these methods and unwittingly undermine the quality of a text's qualitative sections. A qualitatively trained author, who has been involved in writing each chapter, may better motivate the readers to undertake qualitative research or include a qualitative portion in their own investigations.

Among the three of us, we have taught research methods to well over 1,000 management students. As we wrote this text we reviewed each chapter together, debated its content, and revised sentences or sections that any one of us found unclear. We believe that our experience and effort have resulted in a text that engages students and gives them the skills needed to be an effective administrator and community leader.

This book is more than the product of our knowledge, interactions, and teaching. We appreciated the help of Eric Stano and Elizabeth Alimena at Pearson Longman. We especially appreciate their ability to recruit reviewers whose excellent suggestions and criticisms were invaluable. We would like to thank Robert Abbey of Troy University, Nancy Basinger of the University of Utah, Lori Brainard of George Washington University, Laura Littlepage of Indiana University-Purdue

University Indianapolis, and David Shetterly of Troy University for their invaluable feedback throughout the development of this text.

If, after you complete your research methods course, you know how to apply what you learned in your career, we will have achieved our goals in writing this text. We have found that conducting research is unbelievably stimulating as we test our ideas and verify our impressions, and then set to work interpreting the information and deciding what to do next. We hope that you have the same experience.

Practical Research Methods for Nonprofit and Public Administrators

Research in Nonprofit and Public Programs: The Basics

Research in Public and Nonprofit Programs

The Basics

What do you think of when you hear the word *research*? Do you imagine a scientist working in a lab, an economist surrounded by computers and graphs, or a scholar surrounded by a pile of books? You are right; each is conducting research. But research goes well beyond the work of a scientist, an economist, or a scholar. Research may describe what you do when you decide what computer to buy, what jobs to apply for, or what music to download. It also describes what effective managers and influential stakeholders do when they monitor a program's performance, determine its effectiveness, and assess the experiences and opinions of its clients. With information gathered from research, managers monitor their programs, publicize their successes, and identify opportunities for program improvement. Stakeholders may use the information to decide what programs to support and what policies to advocate for. Just as you sometimes make decisions, organizations make some decisions quickly with little information. Sometimes, however, decisions are made after gathering as much information as possible and organizing it. This latter approach to research is the subject of this text. We define *research* as the systematic gathering and analysis of empirical information to answer a question.

RESEARCH AND EFFECTIVE MANAGEMENT

Research and effective organizations go hand in hand. They both depend on openness, accountability, and a commitment to learning and change. Openness or transparency is a key organizational and research value, which adds to an organization's reputation. Stakeholders want to know how an organization uses its resources. They want information on what the organization does and what it has achieved. Information on a program's performance, its effectiveness, and how it is perceived all contribute to transparency. You, your staff, or a contractor may conduct research to gather information. Whoever conducts the research is expected to disclose what information they gathered, whom they gathered it from, and how they analyzed it. With full disclosure, management teams have the information they need to decide if and how they can act on the research findings.

Accountability may be thought of as an attribute of transparency. An accountable organization is open and produces evidence that it is meeting stakeholder expectations. Donors, funders, and clients may all want confirmation that a program is well managed and achieving results. You may collect data on a program's clients, activities, services, and results. Information on clients and activities documents that the program reaches the target population and provides appropriate services. You can link the information on clients and activities to data that can be linked to results. From the analysis you can learn which clients benefit most from a program and what services yield the greatest benefits.

In our eyes the most exciting use of research is to aid in organizational learning and change. The learning opportunities seem limitless. A management team can examine data, gain insights about what is working and what isn't, and explore how a program could do better. Team members with a firm grounding in research methods and analysis may avoid overvaluing research findings and learning the wrong things or undervaluing findings and missing an opportunity to learn.

An organization may conduct a study only because someone else wants it. Managers may comply with the requirements and simply supply the requested information. In doing so they may miss an opportunity to design a study that answers valuable questions or to use the findings to improve their organization. One of our objectives is to empower you to design and implement studies and use findings to further an organization's goals or to improve programs. In this text we consider the words *researcher*, *managers*, *administrators*, and *management team* interchangeable. We assume that you will be able to competently and correctly apply the skills presented here no matter what roles you play in your career.

DEFINING THE RESEARCH QUESTION

Whether you want to learn about programs, organizations, or communities you should start with a research question. The *research question* is a question that someone wants answered. It has two characteristics. First, it can only be answered with observable or empirical information. By definition, research involves the study of observable information. Without observable information no research takes place. Second, the research question should have more than one possible answer. Otherwise why spend the time and money to answer it? While research questions are easy to generate, asking an appropriate question depends on identifying exactly why the study is being done.

Whoever wants the research must be clear on why they want the information, when they want it, how they will use it, and what resources they have available to conduct the research and implement the findings. A management team may consider if the research question is related to the organization's mission and what it will do with the findings. Can the answers justify additional resources, help keep existing resources, or improve employee performance? Studies of inconsequential topics or policies resistant to change are likely to be ignored. A record of producing unused studies is unlikely to lead to organizational strength or career success.

In selecting the specific research questions the management team should decide what evidence will provide adequate answers. Management teams want to avoid "shooting ants with an elephant gun." In other words they do not want to develop a complicated study to answer a simple question. Alternatively, they do not want to design a study that overlooks important information and yields superficial, unusable findings. They do not want to plan a study whose requirements exceeds available resources or yields information too late to help make a decision.

Let's consider a community effort to end homelessness. A task force may want to identify gaps in existing services. To get started it may generate a list of research questions: What programs does the community have for people who are homeless? What population does each program serve? What services does each program offer? What is each program's success rate? Each question requires empirical information. As task force members discuss the questions, they may develop more specific questions, for example:

- How many clients do existing programs serve each month? How long does it take a client to find permanent housing? (These questions address performance.)
- What do clients think about the programs and services? (This question addresses client experiences and opinions.)
- Which services are most successful? What clients and programs have the greatest success? (These questions address program effectiveness.)
- Where should new or expanded services be located? What populations are underserved? (These questions address community needs.)

THE BUILDING BLOCKS OF RESEARCH

Once the study's purpose and the research question are defined you can begin your research. In this section we introduce basic research terms that we will use throughout the text.

Variables, Values, and Constants

Variables, values, and constants refer to the basic components needed to answer a research question. *Variables* are observable characteristics that have more than one value. *Values* represent the characteristics of a variable that change. The values may be dichotomous (yes or no), specific categories (Main Street Center, Southside Shelter, Salvation Army), or numbers.

To answer the question "What clients and programs have the greatest success?" you might ask managers of programs that have homeless clients to report the following information on each homeless client:

Variable	Values
Client age	Under 18, 18–21, 22–25, etc.
Has chronic mental illness	Yes, no
Received job counseling	Yes, no
Received mental health treatment	Yes, no
Received substance abuse treatment	Yes, no
Total services received	0, 1, 2, 3
Months homeless	Actual number of months

If a characteristic has only one value, it is a constant. *Constants* do not vary. For example, if all the programs were in Washington County, Washington County would be a constant.

Hypotheses

Variables are linked to other variables to create a hypothesis. A *hypothesis* is a statement that specifies or describes the relationship between two variables in such a way that the relationship can be tested empirically. A clearly written hypothesis helps you decide what data to collect and how to analyze them. We used the above variables to create three hypotheses:

H_1 Clients under the age of 18 are homeless an average of 14 days.
H_2 Younger clients receive more services than older clients.
H_3 The more services clients receive the less time they are homeless.

In H_1 "Clients under the age of 18" is a constant, the only one in the three hypotheses. We could argue that H_1 is not really a hypothesis since it does not test a relationship between two variables. Still, examining only one variable can provide useful information. Programs for adolescents may require specific information about homeless youth—who they are, what services they use, and how long they have been homeless.

For many research efforts comparisons are valuable. Program managers may want to compare their clients with those of other programs. Staff members may want to learn how one variable is linked to another. H_2 compares number of services received by older and younger clients. H_3 compares the number of services received with the time a person is homeless.

A hypothesis should be specific; it should not be vague or ambiguous. Imagine a hypothesis "Services are related to program success." We do not know (1) if the hypothesis is referring to specific services, the number of services, or something else; (2) whether services have a beneficial or detrimental impact on program success; or (3) what is meant by *success*. H_2 could be more specific. You might want to know what ages are represented by "older" and "younger." To see if the number of services varies across several age groups from the youngest to the oldest clients the hypothesis might be stated as "The older the client the fewer services he or she receives."

A hypothesis implies that a change in one variable brings about a change in another variable. The *independent variable* is thought to explain the variations in the characteristic or event of interest. The *dependent variable* represents or measures the characteristic or event the investigators want to explain. For example, in H_3 the independent variable "number of services received" is thought to explain how long a client is homeless. One may visually link the independent and dependent variables with an arrow from the independent variable pointing to the dependent variable.

Independent variable	$\rightarrow$	Dependent variable
Number of services received	$\rightarrow$	Time homeless

A hypothesis implies a cause–effect relationship. You may state a hypothesis as an if–then statement. For example, "If clients are older then they will be homeless longer." The "if" statement contains the independent variable, that is, client age. The "then" statement contains the dependent variable, that is, the length of time homeless. In our example hypotheses, the independent variables are number of services received (H_3) and client age (H_2). The dependent variables are number of services received (H_2) and length of time a person is homeless (H_3).

For a cause–effect relationship to exist three criteria must be satisfied: a statistical relationship must exist between the independent and dependent variables, the independent variable must occur before the dependent variable, and other possible causes must be eliminated. H_2 and H_3 both imply a statistical relationship between the independent variable and the dependent variable. We may assume that the independent variables in H_2 and H_3 occurred before the dependent variable. Eliminating other possible causes, however, is difficult. You cannot assume that age explains why some clients received more services than others or that the number of services received explains why some clients found housing more quickly.

Hypotheses do not have to imply causal relationships. Just knowing a relationship exists may be sufficient. How would you interpret the findings if they supported H_2? You might wonder if programs discriminate against younger clients, if older clients have different needs, if younger clients are less willing to accept services, or if programs have limited capacity and offer more to those who may benefit the most?

You may wonder why we go to the trouble of stating hypotheses and identifying independent and dependent variables at all. First, this step helps clarify our thinking. As you generate and evaluate hypotheses you can ask yourself, "Why do I think this relationship exists?" "Why do I want to know about this relationship?" "Do I think that the independent variable leads to the dependent variable? Why?" Let's consider H_3. While you might assume that the more services received led to more success in finding housing, it is also possible that the longer a person was homeless the more services he or she was offered. So, in stating a hypothesis and designating an independent variable you have really fleshed out the relationship beyond the bare bones of the hypotheses. Second, identifying the hypotheses and the variables helps in communication. An audience can quickly visualize the question you are asking and how you plan to answer it. Third, specifying the variables and the relationship between them guides your analysis as you decide what to study and how to organize your data.

The Direction of Relationships: An important characteristic of a hypothesis is the pattern of the relationships it postulates. The relationship between two variables may take on one of three forms: *direct, inverse,* or *nonlinear.*

Let's consider the link between age and how long a person is homeless. A direct or positive relationship occurs if the older clients are homeless longer than younger clients. An inverse or negative relationship occurs if older clients are homeless for a shorter time than younger clients. Both direct and inverse relationships may be described as "linear," that is, as the independent variable increases the respective dependent variable increases or decreases. Figures 1.1 and 1.2 illustrate a direct and inverse relationship, respectively

A relationship may have a distinctive but nonlinear pattern. An independent variable and dependent variable may vary together and then level off or change

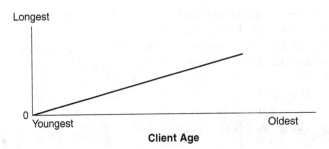

FIGURE 1.1
Illustrating Direct Relationship: Age and Time Homeless

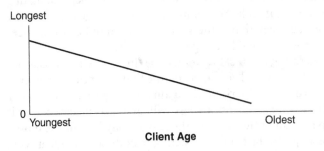

FIGURE 1.2
Illustrating an Inverse Relationship: Age and Time Homeless

direction. Data may show that too much of a good thing can be harmful. For example, too much fertilizer may weaken crops. If the data on fertilizer use and crop yields were plotted, the points might form an upside down U. Each addition of fertilizer may yield better crops, but at some point adding fertilizer will steadily reduce plant quality. The same thing happens to humans. For example, the longer a person stays in a job training program the more likely he or she is to find a job, but at some point the benefit may drop off or actually decrease. Clients who have stayed on "too long" may cease to benefit from the advice they are given. After a period of time some clients may continue to receive training but give up looking for a job. Figure 1.3 illustrates a possible nonlinear relationship between age and number of days homeless. We imagined that among younger and middle-aged persons the relationship between age and time homeless may be direct, but as the person becomes elderly the period of homelessness may decrease.

If possible a hypothesis should specify the direction of a relationship. Postulating the direction will not affect the data you gather or the analysis you do. Your findings may be more valuable if you have laid out a case for the hypothesized directions. Findings that run counter to your expectation may generate thoughtful discussions and new insights.

If an independent variable has no discernible effect on the dependent variable, we say that the two variables do not vary together. If two variables do not vary together, their relationship may be described as *random* or *null*.

FIGURE 1.3
Illustrating a Nonlinear Relationship

Control Variables: In reality a two-variable hypothesis gives limited information. Additional variables, termed *control variables*, may better describe the complexity of why something happens. A control variable may be added to see how it alters the relationship between the independent and dependent variables. A control variable may show that (1) the relationship between the independent and dependent variables is stronger for some values of the control variable than for other values or the hypothesis is supported for some values of the control variable and unsupported for other values, (2) the relationship does not have the same direction for each value of the control variable, (3) the control variable has little or no impact on the relationship, or (4) the original relationship is wrong.

   We have created hypothetical tables to examine the hypothesis that older clients are homeless longer than younger clients. Assume that your data show that younger clients are homeless no more than 30 days and older clients are homeless for more than 30 days. You wonder if older clients are homeless longer because they are more likely to suffer from chronic mental illness. You designate chronic mental illness as the control variable and divide the data into two groups: clients with chronic mental illness and other clients. Tables 1.1 through 1.3 illustrate different impacts of a control variable.

   In Table 1.1, among clients with mental illness older clients are homeless longer. The same relationship is not seen among clients who are not mentally ill: both older and younger clients are homeless for no more than 30 days. In Table 1.2, among clients with mental illness older clients are homeless longer (a direct relationship). Among clients without mental illness younger clients are homeless longer, an inverse

TABLE 1.1

Median Days Homeless of Younger and Older Clients by Presence of Mental Illness: Impact 1

	Median Days Homeless	
Mental Illness	Younger	Older
Yes	<30	>30
No	<30	<30

TABLE 1.2

Median Days Homeless of Younger and Older Clients by Presence of Mental Illness: Impact 2

	Median Days Homeless	
Mental Illness	Younger	Older
Yes	<30	>30
No	>30	<30

TABLE 1.3

Median Days Homeless of Younger and Older Clients by Presence of Mental Illness: Impact 3

	Median Days Homeless	
Mental Illness	Younger	Older
Yes	>30	>30
No	<30	<30

relationship. In Table 1.3, the difference between younger and older clients disappears. From Table 1.3 we would conclude that mental illness, not age, is related to how long a client is homeless. If the original relationship stayed the same, that is, younger clients were homeless for no more than 30 days and older clients were homeless for more than 30 days whether or not they had mental illness, we would conclude that the control variable does not affect the relationship.

Defining Research Subjects

All research depends on subjects. Possible subjects include geographic units, organizations, services, or individuals. The term *unit of analysis* describes what constitutes an individual case. If we collected data on each client of a program to serve homeless persons our unit of analysis would be individuals. To create a list of variables and values you may ask organizations to provide data on individual clients. If our research question was "What are the different client populations, services offered, and success rate of local programs?" the unit of analysis might be program. You might ask of each program for information on the following variables.

Upon analysis you may find that programs with older clients report longer periods of homelessness than programs with younger clients. This finding alone does not indicate that older clients are homeless for a longer period of time because the data do not represent individual clients. You may correctly infer that programs that serve older clients may have unique features. They may offer more long-term treatment programs to address problems associated with old age. They may be

Variable	Value to Report
Percentage of clients with chronic mental illness	Actual percentage
Average age of clients	Median age
Median days clients are homeless	Median number of days
Service provided	
Substance abuse rehab	Yes, no
Job counseling	Yes, no
Medical diagnosis and treatment	Yes, no

more willing to accept clients who have a history of drug abuse or mental illness. As you may have observed, we may be able to aggregate individual data to describe a program. But we cannot do the reverse.

The *population* describes the specific set of subjects we are interested in. The population can be counties in a state, programs in a county, or city residents who are homeless. The respective units of analysis would be counties, programs, and individuals. A *sample* is a subset of the population. We may create a sample to learn about something of interest or we may create a sample to estimate something about the population. For example, to learn about programs serving homeless persons we may decide that we will get the best information by sampling agencies known for excellent services. To learn about what types of programs exist, their clients, and their services we may want to randomly sample from the population of all programs.

THE ORGANIZATION OF THIS TEXT

Sound research skills are important to public and nonprofit organizations and their managers. Effective managers should know how to design and use research to determine how well their organization is doing; how it is seen by their clients, users, and the community; how effective its programs are; and what its community needs are. The public, donors, and legislators, who expect organizations to be transparent and accountable, require such information. Furthermore, the information is needed to help identify problems, design programs, and improve services.

This text is designed to teach you research skills you will use in your career and to give you opportunities to practice applying them to programs. This section of the text introduces basic research concerns. This chapter covered types of research questions and associated terms. Chapters 2 and 3 cover measurement and ethical treatment of research participants, respectively. Effective measurement is critical to any research study. Considering ethical issues that arise in conducting research before beginning a study is essential.

In the later sections we integrate material on methodology and statistics with the types of studies typically conducted by effective managers and programs. These studies may

1. track program performance to provide empirical information on its operations and accomplishments;
2. survey clients, users, and citizens to identify their impressions, experiences, and satisfaction with a program;

3. evaluate a program to demonstrate its effectiveness;
4. assess a community's needs to determine what problems exist and gaps and overlaps in existing services.

Chapters 4 and 5 focus on tracking performance. Chapter 4 presents logic models, which are used to design, implement, and evaluate programs. We introduce them here because they help identify relevant variables. The remainder of the chapter covers how to measure efficiency and track a variable over time. Since performance monitoring typically tracks a single variable Chapter 5 is the logical place for us to discuss graphs and basic statistics.

Chapters 6 through 10 cover surveys. These methods do not apply only to surveys; they play a role in designing other types of studies. Chapters 6 and 7 describe how to select a sample and ask questions, which are basic methodology skills. Chapters 8 through 10 explain how to analyze the data. Chapter 8 focuses on how to describe the data using statistics and graphs. Chapter 9 addresses the question of generalizing from a sample to the population. Chapter 10 explains how to present and analyze information from interviews and open-ended questions.

The next section includes three diverse chapters. Chapter 11 on program evaluation covers research design, a central topic in any introductory methods course. Research designs are of particular importance in collecting and assessing evidence that demonstrate a program's effectiveness. Community needs assessments, the topic of Chapter 12, are not typically covered in an introductory text. We included it here because persons with research training may be expected to help design and conduct a needs assessment. Learning about community needs assessment is consistent with our effort to link research methods with management applications. The chapter on geographic information systems may seem to be an unusual addition to a book on research methods. The use of maps in conducting research is not fully developed, but it is increasing. Maps may play an important role in needs assessment and whenever you think knowing the location of problems, needs, services, or consumers is important. This chapter is intended to help you think about when maps will help you in designing or presenting research.

Chapter 14 covers both oral and written presentations. A study's impact will be enhanced if the findings are presented clearly and effectively. Studies and their findings compete for attention. To illustrate the truth of this, think about how many articles you skim or don't read at all or how many news reports you ignore. Potential users are no different. They may seldom spend time deciphering a complicated report or an unorganized presentation.

A key component of this text is the exercises. Hands-on applications give you immediate feedback about what you understand and what confuses you. We have divided the exercises into the sections Getting Started, Small Group Exercises, and On Your Own. Getting Started asks you to apply the information presented in the chapter to management situations or policy problems. The small group and class exercises mirror what happens in most research projects, that is, each individual participant has ideas about what the research question is, why it was asked, and how to answer it. The exercises in this section require you to develop important skills. You will practice explaining research concepts, presenting your ideas, exchanging ideas with others, and resolving differences. You will find that as you try to explain your ideas, listen to the ideas of others, and incorporate different insights you will understand the methodology better. Through the interaction you should propose an answer that is better than any one person's initial response. The On Your Own exercise is included in most chapters. If you are working or have an internship, you will find this section of the exercise useful as it guides you through the steps required to apply the methodology and to learn more about research practices in the field.

CHAPTER 1 EXERCISES

This section presents a scenario, Sarah's Story, and asks a series of questions that require you to apply basic concepts and debate how to design a useful study.

This section has two exercises, which should develop your skills in linking a research question to stakeholders' needs.

- Exercise 1.1. Learning from Sarah's Story presents a scenario and asks a series of questions that require you to apply basic concepts and debate how to design a useful study.
- Exercises 1.2. On Your Own asks you to identify a relevant research question for your agency (where you work or have an internship) and to consider the constraints and opportunities for conducting a useful study.

Exercise 1.1   Learning from Sarah's story

Scenario

In 1978 the *New Yorker* published a story about a young woman, Sarah, whose first job after college was to research public assistance programs. We have embellished Sarah's story to point out the problems that occur when a researcher starts with no clear direction and to allow you to practice selecting a research question and clarifying the research components.

Sarah's Story (retold more than 30 years later)

Immediately after graduation Sarah applied for a job with a statewide community action agency. The successful candidate would conduct research on the state's public assistance programs. The position was funded for 12 months. Sarah was hired.

Sarah's job interview was short and provided little guidance. The executive director told her not to spend too much time in the library. Rather she should explore and interpret what she saw when she visited the county programs and met county residents who were receiving assistance. She was to submit quarterly reports; her final report would be due at the end of the year.

During the first quarter Sarah drove throughout the state and talked to hundreds of clients in 14 counties. For her first quarterly report she compiled their stories. The executive director called her in and sharply criticized her report. Sarah had totally missed the questions she was expected to answer. The executive director expected a report that the agency could use to advocate with the state legislature.

Section A: Getting Started

1. During the initial interview Sarah and the executive director could have explored the research questions that Sarah was expected to answer.
   a. Suggest a research question focused on
      i. agency performance
      ii. client experiences
      iii. program evaluation
      iv. assessing community needs?
   b. Which question do you think would yield the most valuable information? Why?
2. After Sarah was hired she should have met with agency stakeholders to hear what research questions they would like her to answer.
   a. Write three specific research questions that you can imagine stakeholders would want answered and briefly explain how the information could be used.
   b. What resources will be needed to answer these questions? to implement the findings?
3. Identify how the following could constrain the proposed study.
   a. Personnel requirements
   b. Costs
   c. Time requirements
4. Carry out the following assignments to create and describe a hypothesis and variables.
   a. Write a hypothesis that Sarah could test.
   b. Identify each variable in the hypothesis, state at least two values for each variable, and identify type of each variable (independent, dependent).
   c. State why [name of independent variable] is the independent variable in the hypotheses.
   d. What is the direction of the hypothesis?
   e. Suggest a control variable Sarah might use to further study the hypothesis.
   f. How do you anticipate the control variable affecting the hypothesis?
5. What unit of analysis should Sarah choose to test her hypothesis?
6. What population would Sarah want to generalize to?

Section B: Small Group Exercise

1. Form a group with three to five classmates who suggested a similar research question and (a) consolidate your research questions to state the question that will guide the study, (b) prepare a statement suggesting the value of answering this question, (c) identify types of costs associated with conducting the study and implementing its findings, and (d) identify possible constraints.
2. A spokesperson for each group will present its proposal. After each presentation the class should discuss and evaluate the value of answering the proposed research question and the identified costs and constraints.
3. After the presentations the class as a whole should rank the proposals and decide which one Sarah should pursue.

Exercise 1.2   On Your Own

As part of your job or an internship you may be asked to design a research study. Before expending the time, money, and other resources on a research study you should find the answers to the following questions about it.

1. Why does the management team want the study done?
2. When the study must be completed by [date]?
3. What the findings of the study will be used for?.
4. What resources have been set aside for this study?
5. How does the study advance the mission and objectives of the organization?
6. How can the findings of this study be used to [consider all that apply]
   a. justify additional resources?
   b. maintain existing resources?
   c. monitor a program?
   d. improve employee performance?
   e. improve program implementation?
   f. improve program outcomes?
   g. identify client needs?
7. Identify how the following may constrain this study.
   a. Personnel
   b. Finances
   c. Time
   d. Public opinion
   e. Political realities
8. How will program staff be involved in developing the research question?

Measurement

We live in a world where information is being increasingly quantified. Reviewers rate movies with stars, newspapers report the cost of living in the "world's most expensive cities," and international watchdog organizations rank countries from the most corrupt to the least corrupt. We refer to this process of quantification as measurement. Consider the importance of measurement both in everyday life and to organizations. People use reviewer ratings to decide what movie to see. Tourists use cost-of-living reports to decide what cities to visit. Businesses use information on corruption to decide what international businesses to engage as partners. Organizations use data to identify community needs, to track performance, to evaluate programs, or to learn about clients. An organization may report data to influence policy decisions or to demonstrate accountability. In some cases, such as identifying community needs, the organization may use existing data. In other cases it will develop its own measures and collect the data.

In this chapter we take you through the steps required to decide what you want to measure and determine the quality of measures you use. Later chapters build on this knowledge. Chapter 4 introduces performance monitoring; you will learn strategies for identifying measures to describe individual and organizational performance. Chapter 5 has a brief section on levels of measurement, which is linked to question wording and the choice of statistics. In Chapter 7 we turn to writing questions and questionnaires. Whether you conduct a survey or design forms the specific wording of questions and items directly affect the quality of your measures and the value of your data.

CONCEPTUAL AND OPERATIONAL DEFINITIONS

As we begin to develop measures we may label a variable as a *concept* and the statement that describes what we mean by the concept as a *conceptual definition*. Let's begin with a homey example. What defines a "good restaurant"? Is it one that serves large portions for a relatively low price? Is it one that has interesting, even cutting-edge, food? Or do you look for atmosphere? If you think about your experiences in asking for restaurant recommendations you will understand important characteristics of a conceptual definition. That is, there is no one correct definition of a concept. Whether a definition is appropriate or not largely depends on the users and how they plan to use the information.

The first step is to learn how critical stakeholders define what will be measured and how they plan to use the information. If an adult literacy program manager asked you to help document its effectiveness you might first search for a credible definition of adult literacy. National assessments of literacy use the following conceptual definition: "The ability to use printed and written information to function in society, to achieve one's goals, and to develop one's knowledge and potential," including the ability to read prose, handle forms, and perform arithmetic calculations.[1] The measures based on this definition have been used to track adult literacy in the United States and other countries, but this definition may be of limited use to a specific literacy program. Next you might meet with program staff and ask them what they want to achieve. You could ask staff members how they define literacy, how program activities should add to client literacy, and how they know that a program has been successful. You might ask them how their definitions compare with what clients, funders, and donors expect the program to achieve. Not all definitions are relevant to your study. Definitions, such as those used in national assessments, may help the staff organize its thinking, but they may not measure what a specific program is trying to achieve.

Assume that the staff agrees that they define literacy as "the parents' ability to read to their children and engage them in conversations about the stories." You would then find a way to measure the parents' ability to read the stories, understand the stories, and talk about the stories. Let's just focus on the ability to read stories and understand them. You might present the clients with a few very short stories, ask them to read the stories aloud, and ask a few questions about each story. You would develop a guide to score each client's ability to read the story and answer the questions. The short stories, the questions asked, and the scoring guide constitute an operational definition of literacy. In other words, the operational definition gives you the complete picture of how clients' literacy was determined. The stories included in the operational definition may seem appropriate given the criteria of word length and sentence structure. But a problem occurs if the stories are about unfamiliar objects. A story about farm life may confuse readers with no knowledge of farm buildings, crops, and various animals. To avoid such problems you must conduct a pretest. You should ask persons similar to the program's clients to read the stories and answer the questions. The importance of conducting a pretest is a message that can't be repeated often enough.

Honestly, in the real world, you will seldom encounter the term *conceptual definition*, but you should not ignore its importance. It is the same as asking, "What do we mean by X [the concept or term of interest]?" Too often, once an operational definition is stated people focus on its technical details. They may labor over the wording of the questions and responses. They may not stop to consider if they are measuring what they want to measure.

Let's work through another example focusing on how an organization can evaluate its volunteer orientation program. First, the staff should decide what it

---

[1]This definition is used by the National Assessment of Adult Literacy, an assessment conducted by the National Center for Educational Statistics. For more information go to http://nces.ed.gov/naal/.

wants to accomplish. Should the orientation introduce potential volunteers to the organization, its mission, its history, and its values? Will participants be trained to carry out specific tasks? Are participants expected to sign up as volunteers and recruit other volunteers? Once the goals are decided they are incorporated into a conceptual definition. The conceptual definition serves as a blueprint for deciding what to ask participants at the end of an orientation. The shaded box below uses orientation quality as an example of a conceptual definition and an operational definition and to illustrate their relationship.

Measuring the Quality of Volunteer Orientation

Variable: Perceived Quality of Volunteer Orientation

Conceptual Definition: Orientation attendees understand the organization, value its services, are motivated to volunteer, and feel prepared to work with clients.

Operational Definition, Part 1.

Ask each participant to fill out a form that includes the following items:

For each statement indicate the response that describes your opinion:

1. I understand the mission of [name of organization].
    Responses: Strongly agree, agree, neither agree nor disagree, disagree, strongly disagree
2. I can describe the services [name of organization] offers.
    Responses: Strongly agree, agree, neither agree nor disagree, disagree, strongly disagree
3. I feel qualified to refer clients to [name of organization].
    Responses: Strongly agree, agree, neither agree nor disagree, disagree, strongly disagree

4. I am comfortable with recruiting others to volunteer with [name of organization].
    Responses: Strongly agree, agree, neither agree nor disagree, disagree, strongly disagree
5. I have signed up to be a volunteer with [name of organization].
    Responses: Strongly agree, agree, neither agree nor disagree, disagree, strongly disagree
6. I feel uncomfortable about working with clients.
    Responses: Strongly agree, agree, neither agree nor disagree, disagree, strongly disagree

Operational Definition, Part II

Responses to Questions 1 through 5 are scored 5 = strongly agree, 4 = agree, 3 = neither agree nor disagree, 2 = disagree, 1= strongly disagree. Question 5 uses the opposite scoring: 1 = strongly agree through 5 = strongly disagree. The sum of the six questions represents the perceived quality of orientation. The scores can range from 30 (the highest quality) to 6 (the lowest).

An operational definition determines what we actually measure. In the previous example we learned the perceptions of people who attended orientation. We do not learn if their understanding of the organization and its programs is accurate. We do not know if they are really prepared (or unprepared) to work with clients. We may not know if they actually signed up or actually volunteered. To get that information we would use a different operational definition. For some items we might ask participants specific questions about the organization, its services, and volunteer tasks.

Let's stop here for a word of caution about the pragmatism of operational definitions. You may be tempted to use questions to evaluate volunteer orientation programs that you find on the Internet or obtain from a friend. But a measure that has been designed to assess a general orientation for a large nonprofit with diverse programs, such as the Red Cross or YMCA, might not be appropriate to assess an orientation for volunteers who will work with survivors of domestic violence. Even if a measure has documentation establishing its quality, this does not mean it is the right measure for your study. You need to consider your purpose and the characteristics of the individuals or organizations who will supply the requested information.

Reliability

Before you put a question on a survey or interview guide you should establish its quality. First, you may ask, "Is the measured difference between subjects a real difference?" or "Did measured changes over time really occur?" These questions address the reliability of a measure. Second, you may ask, "Does this measure actually produce information on the concept or variable of interest?" This question addresses the operational validity of a measure. Third, you may ask, "Is this measure sufficiently precise?" This question addresses its sensitivity.

Reliability evaluates the consistency of a measure. Differences over time or between subjects may be due to random error. Random errors are just that, random events or features that affect your findings. Random errors occur because of respondent characteristics, the measure itself, or the process of assigning values. Uninterested or distracted respondents introduce errors when they answer questions rapidly without listening or thinking. Respondents may be inconsistent as they answer items with ambiguous terms or inadequate or unsuitable response choices. Raters may be inconsistent in how they assign values or record answers. Random errors cannot be entirely eliminated, but you want to make sure that they are not undermining the value of your data. A measure that yields a large amount of random error should be discarded.

Dimensions of Reliability: Reliability has three dimensions. A reliable measure should have stability, equivalence, and internal consistency. A stable measure yields different results over time only if the phenomenon being measured has changed. Consider choosing scholarship recipients. To ease its work and assure a fair selection process, the selection committee may create a measure to rate the applications. For each application a rater adds up the values of separate items that measure experience, leadership, achievements, and potential future contributions. The assigned values or ratings should not change with the rater's mood, fatigue, or degree of attention. To make sure the ratings are consistent the raters can re-rate a random sample of applications. If their original ratings and second ratings are the same or nearly the same, the measure may be deemed stable.

An equivalent measure yields different results only if the differences between subjects are real differences. For example, if you and some colleagues use the same measure all of you should assign the same value to the same person or case. If several people are to rate the scholarship applicants they should first rate a sample of applications. The sample should represent a variety of applications, including applicants with diverse backgrounds or applications with wordy answers. If the raters give the same or nearly the same ratings to each application we can assume that the differences between subjects are real differences and not due to differences in the raters. If the raters give different scores the measure has to be examined to identify and resolve problems.

Equivalence also applies if two or more versions of a measure are used. Consider written driving tests. If each applicant takes the same test, cheating becomes a problem, so numerous versions are created. An equivalent measure would ensure that comparable test takers receive the same score no matter which test version they take. Developing alternate versions, however, is time consuming and rarely done except when designing a test for large numbers of people.

Internal consistency applies to measures with multiple items. It establishes if the items are empirically related to each other. A simple example may help explain internal consistency. Imagine a measure of arithmetic skills with five multiple-choice questions. The questions are

1. $35 + 83 = ?$
2. $83 - 35 = ?$
3. $83/23 = ?$
4. $1.83 \times 23 = ?$
5. $5e^3 = ?$

Let us assume that items 1 to 4 relate to arithmetical ability and that item 5 does not. Respondents who have good computational skills may easily answer the first four items, and guess the answer to item 5. Their guesses represent random error. If item 5 was dropped the measure would be more internally consistent.

Qualitative Evidence of Reliability: The next question to ask yourself is, "How much random error should I tolerate?" Think about the examples we have used in this chapter—adult literacy, quality of volunteer orientation, and scholarship ratings. Which measure should have the least random error? Why did you choose this measure? To decide how much random error to tolerate, you should consider its consequences. Among our three examples we would want to have the least error in choosing scholarship recipients, since the reliability of the measure will affect who receives the scholarship and who doesn't. The decisions based on the other two measures will neither benefit nor harm an individual. Now think about driving tests. Given the large number of drivers and the potential of an unsafe driver to cause great harm to others the need for minimum random error is even greater.

To estimate reliability start with a qualitative approach. Qualitative methods do not precisely estimate the amount of random error. Hence, you may underestimate their value. However, at a relatively low cost, they can identify problems that will thoroughly discredit a poor measure. You should review a measure's operational definition to decide whether

- terms have been defined precisely;
- ambiguous items or terms have been eliminated;
- information is accessible to respondents;
- multiple-choice responses cover all probable responses;
- directions are clear and easy to follow.

A few examples will demonstrate the problems introduced if a measure lacks one or more of these characteristics. Recall that an unreliable measure does not detect actual differences between subjects or in the same subject over time. Before you ask individuals if they are homeless or ask an agency how many of its clients are homeless you need to define *homeless*. Homeless persons may include persons living on the streets, in a shelter, doubled up with relatives or friends, or couch surfing. Unless you have defined the term some people who are living with relatives "until something comes along" will say they are homeless and others won't. The same may be true of young people who are sleeping on friends' sofas. If you do not define what you mean by *homeless*, differences between people who say they are homeless and those who don't may not be actual differences but differences in how they interpreted the question.

Gathering data about *family size* or *earnings* may seem straightforward, but both are fraught with ambiguities. If you ask "What is the size of your family?" you may be asked if half-brothers and sisters should be counted. What about divorced parents or siblings who live in different households? What about relatives or their partners who are part of the household? Similarly, asking members of a diverse population how much they earn is tricky. If a time period is not indicated respondents may report hourly, weekly, monthly, or yearly earnings. A person who reports "$12,000" may be reporting her monthly or her annual earnings. On the other hand seasonal workers and self-employed persons may not know how much they earn in a year. Similarly, the accuracy of estimates of family income may be wildly inaccurate depending on which family member is asked.

To reduce the random errors and increase reliability you should review the definition of terms, the clarity of items, the accessibility of information, and the appropriateness of response choices. Your review of the operational definition should involve comments from potential respondents. Otherwise you risk falsely assuming that actual respondents will interpret the items the same way you do. Remember, don't be presumptuous. Spell out what you want to know.

Additionally, the reliability of measures also depends on interviewers and data entry staff. They must be trained and supervised to minimize inconsistent decisions. For example, all interviewers should define *family* the same way and use the same procedures for resolving problems of inaccessible information or ambiguous questions. Similarly, staff entering data on a spreadsheet may be uncertain how to handle forms on which respondents checked more than one response, embellished a response category with their own comments, or added an alternative response. If individuals decide on a case-by-case basis how to handle ambiguous responses, the decisions may be inconsistent and the data less reliable.

Quantitative Evidence of Reliability: All measures should receive a qualitative review before they are used to collect data. In addition mathematical procedures may be used to estimate reliability. Specific tests estimate the measurement error associated with stability, equivalence, and internal consistency. We will limit our discussion to tests that you may find useful and easy to implement and ones that are commonly reported in research articles. If you plan to construct or work extensively with job tests, achievement tests, or personality tests you need to become familiar with other methods for mathematically estimating reliability.

Test–retest establishes a measure's stability. You should conduct a test-retest before collecting data. You need to collect the data at two points in time and you should expect that what is being measured will not have changed. Let's consider the scholarship selection committee. To establish a measure's stability you may ask the scholarship selection committee to rerate some of the applications. If the raters give the same scores, you cannot automatically assume that the rating measure is reliable. You have to verify that raters did not remember and reproduce their original ratings. If the scores are different you may assume that the instrument is unreliable. After the first round of ratings the raters may have discussed the measure and changed how they valued the various items. The information on the applications will not have changed. The perceptions and ratings of the raters changed. Using test-retest to establish stability is difficult insofar as an instrument itself may be the source of a change in values.

Inter-rater reliability establishes the equivalence of a measure. Inter-rater reliability should be established if two or more people are collecting data. Here, observers apply the measure to the same phenomena and independently record the scores; next, their scores are compared. Our example of several people rating the scholarship applicants and comparing their ratings illustrates inter-rater reliability. You may apply inter-rater reliability to staff training. Imagine training housing inspectors. Trainees may inspect and rate a sample of dwellings. The trainee may be approved to work once her ratings agree with the ratings of experienced inspectors.

Tests of internal consistency establish the homogeneity of a measure. Internal consistency is used only if a measure consists of several items. The key statistic to assess internal consistency is Cronbach's alpha. The closer alpha is to 1.0 the more internally consistent the measure. If alpha is close to 0.0, the items are unrelated to one another. As is true with many statistics, there are no set criteria for an acceptable alpha. When a research report indicates a measure's reliability it normally refers to a test of internal consistency.

At a minimum you want to verify that the data are free of gross errors. To do so, start with a qualitative review of a measure's operational definition. The review will identify serious threats to reliability. Depending on how a measure will be used you may consider a quantitative test as well. On reading research reports you are most likely to see references to inter-rater reliability or internal consistency.

Operational Validity

Operational validity produces evidence that a measure is correctly named and accurately describes what it is measuring. Let's consider poverty. In the United States a household is under the poverty threshold if its income is less than three times the cost of a minimally adequate diet. The multiplier of three comes from a 1955 finding that a typical family spent one-third of its income on food. The challenge of measuring poverty illustrates several important points. First, operational validity depends on planned use. A measure of poverty may be used to track changes in poverty over time, determine who should receive financial assistance, or describe the effect of poverty on individuals and their community. The poverty threshold has the advantage of being reliable, and changes from year to year may be assumed to be real changes. It may not, however, be a fair criterion to decide if a household needs financial assistance.

Second, operational validity depends on the content of the measure and relevance. A measure of poverty may take into account the cost of food, housing, and medical care. Other possible content includes the cost of child care and transportation. The relevance depends on how the data will be used and interpreted. Over time the poverty threshold has become less relevant, because food consumes far less than one-third of most family budgets.

Third, operational validity always involves judgment. Some people may argue that a poverty measure should consider only if households can meet the most basic needs. Others may argue that education, child care, or transportation should be included, because without them, families may stay impoverished. It is not possible to prove which is the more valid. Judgment also comes into play when measures are modified. Users may have to decide whether to sacrifice comparability or reliability in order to have a more relevant measure.

An important measure, one that will impact decisions, inevitably spawns disagreement about its validity.[2] Should a measure of poverty account for entitlements, such as food stamps? Is a measure of good government biased toward the values of businesses? Does gross domestic product (GDP) measure progress if it ignores pollution and social inequities?

Designing a measure takes considerable work, skill, and thought. While you cannot prove that a measure is operationally valid, you can produce evidence that its content is representative and it is relevant. In the following sections, we discuss the evidence based on a measure's content, its relationship to other measures, and its consequences.

Evidence Based on Content: Establishing evidence that a measure is valid begins with the conceptual definition. The appropriate content may emerge later in discussions with stakeholders about what the concept means to them and how they plan to use the data. The next step is to design an operational definition that represents the conceptual definition. Let's go back to our example of the quality of volunteer orientation. The conceptual definition was that attendees would understand the organization, value its services, be motivated to volunteer, and feel prepared to work with clients. We chose six items to measure the quality of orientation.

---

[2]You can find information on the Internet about many types of measures and indicators. For example, if you want to get an evaluation of indicators of volunteerism, you could enter "volunteerism indicator critique" into a search engine. You will find sources that present a critique that will add to your understanding of measurement and the construction of measures. For this paragraph, we relied on the following articles and Web sites (all accessed on June 7, 2010): Tina Soreide, "Is it right to rank? Limitations, implications and potential improvements of corruption indices," Chr. Michelsen Institute, Norway, http://www.cmi.no/publications/publication/?1973=is-it-right-to-rank-limitations; Daniel Kaufmann, Aart Kraay, and Massimo Masruzzi, "The Worldwide Governance Indicators Project: Answering the Critics," The World Bank, http://www-wds.worldbank.org/external/default/WDS ContentServer/IW3P/IB/2007/02/23/000016406_20070223093027/Rendered/INDEX/wps4149.txt; "Measuring Progress," Friends of the Earth, http://www.foe.co.uk/campaigns/sustainable_development/progress/replace.html

For each item the respondent used a scale ranging from "strongly agree" to "strongly disagree."

Understand the organization:
   I understand the mission of [name of organization].
   I can describe the services [name of organization] offers.

Value its services:
   I feel qualified to refer clients to [name of organization].
   I am comfortable with recruiting other volunteers to [name of organization].

Motivated to volunteer
   I have signed up to be a volunteer with [name of organization].

Prepared to work with clients
   I feel uncomfortable about working with clients.

You have to compile a set of items that you believe adequately captures the conceptual definition. We will leave it to you to decide if the six items listed here are representative. It is a matter of judgment. To ask about every aspect of orientation may require many more questions—far more questions than respondents would want to answer or staff would want to analyze.

As we noted earlier, the operational definition measures attendees' perception and does not confirm that they understand the organization's mission and its services, are capable of working with clients, or are willing to actually volunteer. Thus, you cannot assume that attendees' perceptions are a valid measure of the actual quality of orientation.

A measure's relevance is directly linked to its intended use. The six items provide a snapshot of attendees' opinions about orientation. They may identify weak spots in orientation. However, the items may be inadequate in a different context for example, if the information is meant to guide revisions in the orientation program—keeping what works and dropping what is ineffective. Also, the decision to measure perception rather than objective measures of quality may depend on how the data will be used and interpreted. Creating and administering an objective measure may seem infeasible or unnecessary for its intended use.

Evidence Based on Relationship to Other Variables: Reviewing a measure's content does not tell the whole story. You will learn a lot if you compare the data from one measure with data from another measure. We will refer to other measures as *criteria*. You might select a similar validated measure as a criterion. For example, an adult literacy program manager may compare instructors' assessments of the reading ability of individual students with their performance on standardized reading tests. If the instructors' assessment agrees with the standardized test scores the program can forego the cost of obtaining and administering standardized tests. The program will have evidence that instructor assessments are a valid measure of client reading ability.

You may select independent assessments as criteria that examine self-assessments and perceptions. Self-reports of health may be compared with a physical exam or a person's medical records. Employee assessments of their job performance may be compared with agency records. Client assessment of their reading ability may be compared with their instructors' assessments.

You may select criteria that are logically linked to the measure. You would expect people who gave orientation a high rating to volunteer more than those who didn't. Likewise, you would expect people who are categorized as poor to have a less nutritious diet than other people. You would expect agencies with high client satisfaction to perform better than ones with low client satisfaction.

Finding a weak relationship between a measure and a criterion may be disconcerting, but it may start a valuable exploration of what information a measure is providing and what it is not providing. We may find that people say that they are healthier than what their medical records indicate. Because we cannot argue that perceived health validly measures a person's health, the information is valuable. A measure of perceived health can help explain what motivates people to seek medical care or why they ignore serious symptoms.

Another type of criteria is a future outcome. Consider selecting scholarship applicants. Assume that a scholarship's purpose was to train future leaders. The committee chose and measured items that it believed would identify applicants who use the training and become leaders. To estimate if their selection criteria were appropriate the committee might track the career outcomes of applicants. Similarly, organizations would want to choose employees who will perform well in their jobs. From time to time they may review their selection measures to see if they still capture the quality of employee performance. From examining the outcomes of scholarship selection and employee performance we can learn if people who were expected to be successful were successful. If the criterion includes only qualified and selected people, we will not know if applicants who were rejected based on the criterion would have been equally good or better.

The Consequences of Applying a Measure: Traditionally, operational validity asks if you are measuring what you intend to measure. A closely related question is what are the consequences of applying a measure. The more importance stakeholders place on the data a measure produces the more they should consider its consequences. Administrators and policy makers assume positive benefits when they first adopt a measure. They should confirm that the positive benefits were realized and identify any unintended consequences. The No Child Left Behind Act is a vivid example of the consequence of measures. The act was intended to ensure that all American school children are proficient in mathematics and reading. It called for annual testing of children in third through eighth grades. If a school did not meet statewide performance standards, it was expected to take remedial action. The policy makers assumed that the tests would motivate schools to hire qualified teachers, adopt effective teaching practices, and assist children in danger of failing. Critics have argued that the act's emphasis on test scores has led states to create easier tests and increase the pass rate, shifted the curriculum away from developing critical thinking, discouraged teachers from teaching in low-achieving schools, and encouraged low-achieving students to drop out.[3] This example reminds us that what is being measured can affect behavior. Sometimes the impact is not what we expect or want.

---

[3]L. Darling-Hammond, "Evaluating 'No Child Left Behind,' " *The Nation*, May 2, 2007. Posted at http://www.thenation.com/doc/20070521/darling-hammond (accessed December 7, 2009); J. E. Ryan, "The Perverse Incentives of the No Child Left Behind Act," *New York University Law Review*, June, 2004.

Sensitivity

The *sensitivity* of a measure refers to its precision or calibration. A sensitive measure has sufficient values to detect relevant variations among respondents; the degree of variation captured by a measure should be appropriate to the purpose of the study. Measures that are reliable may not necessarily detect important differences. Consider a salary survey. Suppose employees were asked the following question:

What is your annual salary for this year? (check appropriate category)

_____ Less than $25,000

_____ $25,000 to $39,999

_____ $40,000 to $54,999

_____ $55,000 to $69,999

_____ $70,000 to $84,999

_____ $85,000 or more

These categories may be adequate to summarize the earnings of all employees, but not the salaries of senior managers. If most top managers earn at least $85,000, all the responses will fall in one category. You may not be able to learn if salary is related to performance.

The sensitivity of a measure may depend on the homogeneity of respondents. For example, a job-satisfaction measure developed for organizations employing unskilled and skilled laborers; clerical workers; and technical, administrative, and professional staff may be a poor choice for an organization largely made up of professionals. If individual differences are of interest, the measure would not be sufficiently sensitive to compare differences among employees in the more homogeneous group.

CONCLUDING OBSERVATIONS

No measure can describe fully a concept of interest. At best a measure allows us to estimate what program clients achieve, volunteer and employee productivity, or the number of poor. The information a measure provides has great value, but you should also recognize each measure's limitations.

Measurement begins with a conceptual definition. The conceptual definition indicates what stakeholders mean by a concept and how they plan to use the data. The conceptual definition serves as a blueprint for the operational definition, which details exactly how a concept or variable is measured and its values are determined. The conceptual definition will be particularly important to decide if the measure's content is representative and valid.

You should establish the reliability, operational validity, and sensitivity of all measures before data collection begins. Reliable measures allow you to conclude that differences between subjects or over time are real differences, and not due to the measure or the measuring process. A qualitative review of a proposed operational definition will markedly improve reliability and may be adequate. You should verify that directions are clear and easy to follow, that items are clearly defined, that given responses cover all likely responses, and that respondents can provide the requested information. People responsible for collecting or processing data should be trained to avoid inconsistencies. If knowing and limiting the amount of random error are important, you should use mathematical procedures to estimate the amount of error. Unreliable data should be discarded.

A reliable measure is not necessarily operationally valid. A valid measure measures the concept of interest. Evidence based on a measure's content documents that the operational definition is relevant and representative. Criterion-based evidence empirically establishes the relationship between a measure and a criterion.

The criterion may be a similar measure, an alternate measure, or a future outcome. Evidence may also be gathered to document the consequences of implementing a measure. The evidence validating a measure informs and supports the judgment of data users; it does not replace their judgment. A sensitive measure sufficiently distinguishes cases from each other so that they can be compared.

RECOMMENDED RESOURCES

More detailed discussions on measurement, especially quantitative evidence of reliability, can be found in texts on educational measurement or psychometrics. A recommended text is Susana Urbina, *Essentials of Psychological Testing* (New York: John Wiley & Sons, 2004), which is a basic, accessible resource on measuring, including knowledge, abilities, attitudes, and opinions.

A standard reference on measurement is *Standards for Educational and Psychological Testing* (Washington, D.C.: American Educational Research Association,

1999). The standards were created by a joint committee of the American Educational Research Association (AERA), American Psychological Association, and National Council of Measurement in Education and approved by their respective governing bodies. The AERA endorsement stated, "We believe . . . the *Standards* to represent the current consensus among recognized professional[s] regarding expected measurement practice. Developers, sponsors, publishers, and users of tests should observe these standards" (*Standards*, viii).

CHAPTER 2   EXERCISES

There are three separate exercises for this chapter. Each exercise develops your competence in interpreting and applying measurement concepts.

- Exercise 2.1 Good Nutrition Survey focuses on the strategies for designing reliable and operationally valid measures.
- Exercise 2.2 Living Wage: From Idea to Measure focuses on the concept of living wage. The exercise suggests the relationship between measurement and politics, and how a concept is measured and how the data will be used. The exercise implies the complexity of developing a sound operational definition.
- Exercise 2.3 Selecting Job Applicants focuses on employee hiring and implementing strategies to develop and assess the reliability and operational validity of the selection process.
- Exercise 2.4 On Your Own asks you to design and assess a relevant measure for your agency (where you work or have an internship).

EXERCISE 2.1   Good Nutrition Survey

Scenario

A nutritional council has as its mission to improve public knowledge of nutrition and advocate for healthier diets. You have been asked to help the council design a self-administered survey to measure an individual's knowledge of sound nutrition and eating habits.

Section A: Getting Started

1. Why should the council develop a conceptual definition?
2. Find three conceptual definitions of good nutrition.
3. Which of the three definitions seems most appropriate for the nutritional council to use? Justify your choice.
4. Based on the conceptual definition write an operational definition.
5. What steps would you take to develop evidence that your operational definition is reliable?
6. Should you focus on the qualitative evidence of reliability prior to implementing any procedure to establish quantitative evidence? Justify your answer.
7. What evidence would you cite to argue that your operational definition is operationally valid?
8. Consider the appropriateness of using your operational definition to measure the nutrition levels of
   a. a predominately Hispanic community.
   b. pregnant women.
   c. low income children.
9. In the course of writing surveys, you may become concerned with the reliability or operational validity of specific questions. Comment on each of these proposed questions.
   a. Does your family eat a healthy diet?
   b. How many servings of protein do you eat daily?
   c. How many times did you eat breakfast in the past 7 days?
   d. How many calories do you consume in a typical day?
10. Long surveys often have a low response rate. Would you argue that asking only one or two questions is sufficient to measure an individual's knowledge of good nutrition? Justify your answer.

Section B: Class Discussion

1. Based on the conceptual definitions that class members found, consider which definition seems most appropriate to
   a. design an educational campaign, Eat Smart.
   b. design sample menus and recipes for council publications and fliers.
   c. create a dietary checklist for medical personnel to include in a physical exam.
2. Choose one of the operational definitions identified and use the qualitative indicators of reliability to assess how well it would work for a
   a. telephone survey of the general population.
   b. self-administered survey of parents of children who receive subsidized meals.
   c. self-administered survey of elderly persons who eat one meal a day in senior centers.
3. What evidence suggests that the operational definition is operationally valid?
4. Questions may focus on respondents' perception of the quality of diet or they may ask for more specific information, such as, listing foods actually eaten in the past 48 hours. Assess the benefits and drawbacks of asking
   a. questions about what people believe regarding different types of food.
   b. questions about what foods people eat.

EXERCISE 2.2   Living Wage: From Idea to Measure

You have been asked to serve on a community task force to consider whether it should adopt a livable wage ordinance.

Section A: Getting Started

1. What does the term *livable wage* mean to you?
2. Search the Web to identify a discussion of the livable wage: what it means, its desirability, the difficulty in determining what constitutes a livable wage. What are the major measurement issues that the discussion focused on?
3. Do a Web search and find two more conceptual definitions of livable wage.
4. Deciding on an appropriate conceptual definition is a matter of judgment colored by how the data will be used. For each potential user on the following list, identify which definition you think would be preferred and why.
   a. By employers deciding on an acceptable salary for their lowest-paid employees
   b. By an advocacy group lobbying for an increase in the minimum wage
   c. By a public agency determining who is eligible for services, such as financial assistance, food subsidies, and medical services.
5. Respond to the following items to develop an operational definition for "livable wage."
   a. Identify the components you would include in an operational definition of a livable wage.
   b. In your operational definition, will some items receive more weight than others or will all items be weighted equally? Explain.
   c. Do you anticipate that the operational definition will work equally well in urban and rural areas? Explain.
6. Consider measuring the cost of rental housing.
   a. Write an operational definition for cost of rental housing.
   b. Would you expect to get more reliable data from landlords or renters? Explain.

Section B: Class Exercise

In assigned groups hold a task force meeting. Discuss the following:

1. Share the conceptual definitions and identify which definition(s) the task force should adopt or present for public debate.
   a. What components of the conceptual definition do group members agree on and disagree on?
   b. Are the reasons group members disagree based on ideology, purpose of the measure, impact of the data, feasibility of developing or implementing an operational definition, or something else?
2. Identify how you would decide what wage would constitute a livable wage in your community.
3. Identify what operational definition you would recommend to identify the number of wage earners in your community who do not earn a livable wage.
4. Based on what you have learned and observed in answering Sections A and B create a list "Lessons Learned about Measuring Concepts."

EXERCISE 2.3 Selecting Job Applicants

Reliability and operational validity apply to many tasks, especially if we want to rank or categorize something. In this exercise we ask you to use your knowledge of measurement and develop a procedure for reviewing job applications for the position of Director, Whitney M. Young Center. After reading the announcement, respond to the items listed after the job announcement to guide you.

Job Announcement: Director

The **Whitney M. Young Center for Urban Leadership (WMYCUL)**, an affiliate of the National Urban League, is a nonprofit educational institute that exists to foster positive social and economic change through effective leadership development. Our mission is to cultivate and enhance the leadership capabilities of individuals and organizations that serve urban communities. Our role is to convene and link those entities to practical leadership development tools and resources to help them address capacity issues and various leadership challenges. Our philosophy is that through the enrichment and/or development of urban leaders, urban communities can be improved and empowered, thereby helping to create a society whereby all people have equal opportunity to be positive contributors.

Position Description
The director coordinates and manages the professional development/leadership training needs of 100-plus affiliates located throughout the United States and the National Office.

Essential Functions
- Oversees professional development/leadership training for the affiliates and National Office. Develops videos, resources guides, handbooks, manuals, and guidelines for the affiliates. Identifies 21st century resources to enhance system-wide capacity.
- Coordinates and monitors training activities, including annual leadership conferences.
- Works with staff to develop strategies to support strategic direction of center
- Performs other duties at assigned

Requirements
- Master's degree in social work, public administration, or related area
- Certified Professional in Learning & Performance (CPLP) a plus

---

- Seven to ten years experience designing training curriculum, program design and execution, or leadership development; experience with national professional training network a plus
- Able to plan, organize, control, delegate, and manage multiple projects simultaneously
- Familiar with nonprofit governances and operations
- Strong analytical, presentation, and facilitation skills. Must have meeting-planning experience. Excellent interpersonal and communication skills.
- Proficiency in Microsoft Office.

To apply, submit resume, cover letter, and writing sample to Human Resources Department, or e-mail recruitment@thisjob.com. Deadline September 7.

The National Urban League is the nation's oldest and largest community-based movement empowering African Americans and other people of color to enter the economic and social mainstream. The National Urban League is an Equal Opportunity Employer.

Section A: Getting Started

1. Why is it important to have a reliable and valid process for selecting which applicants to interview?
2. You plan to create a rating scale, that is, you identify and assign a numeric value to each key requirement. How would you use qualitative methods to make sure that the scale is reliable?
3. How would you use test–retest to establish the scale's reliability?
4. Assess the value of using test–retest to establish the scale's reliability.
5. How would you use inter-rater reliability to establish the scale's reliability?
6. Assess the value of using inter-rater reliability to establish the scale's reliability.
7. Create a plan to establish the scale's reliability, that is, consider qualitative methods, test–retest, and inter-rater reliability. Which strategy(ies) would you use, why, and in what order would you use them?
8. What would you look for to establish the scale's operational validity?
9. If after you complete the first round of interviews you find that none of the interviewed applicants were suitable, would you conclude that the problem was with the reliability or the operational validity of the scale? Justify your answer.
10. Review the job description and identify the requirements that should be included in the scale. Also note if any of the requirements should be given more weight.
11. Examine the following two resumes and create a scale that could be used to rate each resume.

J. Q. Public
123 Main Street
Anytown, Yourstate, USA
(111) 555-1234

Objective

To apply my education and strong work ethic toward a career in the public relations, specifically event planning and program coordination.

Keywords

Event planner, event coordinator, meeting planner, wedding planner, project manager, communications, advertising, public relations, entertainment.

Experience

Promotions Coordinator, Local Radio Station (2007–present) Planned, coordinated and executed over 200 on-site promotions and remotes. Assisted the Marketing Department in developing and executing station contests, promotions and events. Worked with programming department to carry out station programming agenda. Directed promotional street teams to implement on-site promotional campaigns. Created new contests and promotions that drive station-marketing objectives. Engineered remote broadcasts. Assisted in building and maintaining client relationships. Helped to execute many large-scale events including the Annual Ball, Community Jam, and Cultural Fest.

Executive Director, Any Magnet school, Inc. (2004–2007). Planned major special events and lead extracurricular activities. Worked with administration, faculty and student teams to plan homecoming events and fundraising dinners; tracked all milestones and ensure on-time, successful completion. Planned fundraising events such as International Day, including advertising/promotion, decorations, costumes, student presentations and sponsorship. Directed annual African American Heritage Festival, student competition with judges and awards ceremony attended by more than 500 people; program included dinner buffet and presentation of group projects on educational topics.

Project Manager. Human Services Agency, Inc. (1999–2004) Responsible for implementing the Independence Program, a welfare to work program that assisted heads of households transition from welfare to self-sufficiency through employment. Developed relationships with area employers and promote program services. Serve as liaison to collaborating agencies including State Department of Human Resources, City Department of Social Services, and other community based organizations. Hired and supervised a professional and paraprofessional staff of 15. Developed and revised forms, manuals, brochures, and MIS systems. Compiled and provided monthly and quarterly reports to funders and stakeholders.

Program Director. Children's Human Services, Inc. (1994–1999) Responsible for the successful implementation of a childcare center for homeless children with special needs. Managed program budget of $275,000. Supervised a staff professionals and paraprofessionals and college level human service program interns. Coordinated program activities with collaborating agencies for service provision, grant writing, planning, resource sharing, and development. Compiled and provided reports for funders and stakeholders.

Education

MSW and MA in Organizational Leadership, Midstate University (1995).
BA, Communications, Atlantic State University (1990).

Additional Information

Computer skills include Windows, Word, Excel,
PowerPoint and Internet research.

Jane Doe
9876 Elm Street. Springfield, (111) 555-1234

I am interested in a Event Planning position for a corporation or nonprofit organization. In addition to the Bachelors of Science Degree in Hospitality Tourism Management/Business Administration, with an emphasis in Events, Attractions and Conventions Management, I also possess the following qualifications and areas of expertise:

Education

Columbia Southern University: MBA with Concentration in Hospitality and Tourism

University of Delaware: Bachelor of Science Hotel, Restaurant, and Institutional Management

Work Experience

Independent Contractor – Open Events, LLC. 2004–present
- Assist with the planning and management of annual sales meetings for international high-tech companies,
- Work with and coordinate food and beverage staff, audio-visual personnel, travel department, ground transportation, security and labor.
- Oversee the planning and execution of seminars in the US, Caribbean, and Europe
- Perform site research and negotiation, menu selection, registration and travel coordination, shipping and receiving materials between destinations and the on-site management of the entire event program.

Corporate Affairs Director – Telephone company Public Relations 2000–2004

- Established and leveraged strategic relationships with key stakeholders in the Hispanic, Asian, and businesswomen's markets. Successfully positioned these associations to win support for Telephone Co. sales initiatives and public policy.
- Partnered with Supplier Diversity division to ensure engagement with minority suppliers of all groups, i.e., African American, Asian Pacific American, Hispanic, and Women.
- Directed Telephone co. multimillion dollar philanthropic investments with these and other national organizations via the Telephone Co. Foundation and business unit funding. Garnered significant coverage in mainstream as well as Spanish-language and Asian-language media for Telephone Co. grants and sponsorships.
- Secured more than 3,100 constituent letters sent to Congress and several op-eds in support of Telephone Co. public policy positions.

Promotions Coordinator- Local Radio Station 1994-2000

- Planned, coordinated and executed over 200 on-site promotions and remotes.
- Assisted the Marketing Department in developing and executing station contests, promotions and events.
- Worked alongside programming department to carry out station programming agenda.
- Worked in conjunction with promotional street teams to implement on-site promotional campaigns.
- Helped to create new contests and promotions that drive station-marketing objectives.
- Assisted traffic department with entering sales orders, filing, and writing contracts.
- Engineer remote broadcasts.
- Assisted in building and maintaining client relationships.

Other Skills & Experience

Self-starter, goal orientated, assertive, and possess a warm outgoing personality. • Excellent time management skills, detail orientated, strategic thinker and a track record of making sound decisions. • Style which exhibits maturity, high energy, sensitivity, teamwork, and the ability to relate to a wide variety of professionals. • Strong interpersonal communication skills with ability to effectively identify and fulfill customer needs • Fluent in Portuguese and Spanish • Familiar with Mac, PC and Internet platforms, Microsoft Word, Excel and Quicken

12. Now rate both applicants using your scale.

Section B: Class Discussion

1. In small groups review each member's rating scale.
   a. Use the qualitative method to evaluate the reliability of each person's scale reliability. Indicate what changes need to be made to improve any problematic item's reliability.
   b. Assess each scale's operational validity. Taken as a whole does each scale's choice of items seem relevant? Do the items seem representative? What evidence supports your conclusion?
   c. What empirical evidence would you seek to demonstrate that the scale was operationally valid?
2. Choose one of the scales and have each member of the group apply it to the two resumes. To assess the inter-rater reliability of the selected scale. Compare your ratings.
   a. How similar are they?
   b. On which items are your ratings the most different? How can you improve these items?
3. Based on this exercise draft a guide "Suggestions for developing and assessing the reliability and operational validity of measures to select applicants for jobs and promotions." Compare your guidelines with those developed by your classmates.

EXERCISE 2.4: On Your Own

This section is applicable if you are working or have an internship. This section of the exercise guides you through the steps required to design a reliable and operationally valid measure and to observe the value of stakeholder feedback.
   Consider the concepts you are trying to measure. List and prioritize them.

1. From among these concepts select a key concept. Find out how similar organizations define it. Be sure to look at the relevant literature for definitions.
2. Consider how these definitions are similar or different. Is the concept or definition contested? Has it evolved over time? If so, in what ways and what influenced the change?
3. Adapt an existing operational definition for the concept or develop your own.
4. Assess its reliability and operational validity.
5. Review your conceptual and operational definition with relevant colleagues or other relevant stakeholders.
   a. Do they agree that you are measuring what you intend to measure?
   b. Do they agree that your operational definition is both relevant and representative?
   c. Do they raise concerns about the wording of items and the responses, the willingness and ability of respondents to provide the information?

Ethical Treatment of Research Subjects

At some point during the research process investigators may require research subjects, that is, people who will answer surveys, agree to interviews, participate in focus groups, or enroll in a demonstration project. People who agree to answer questions or participate in a study expect to be treated respectfully and ethically; they do not expect to be harmed by merely participating in a study.

Subjects who provide information for community assessments or answer stakeholder surveys are unlikely to experience physical harm or acute psychological distress. Rather the "harm" they experience may be subtle, such as losing some privacy, wasting their time, or experiencing unpleasant emotions such as anger, defensiveness, or distrust. Staff that design, implement, and evaluate a program must be attuned to a study's potential to cause significant harm to participants.

This chapter covers U.S. regulations and professional standards that apply to community and organizational research involving human subjects. To put current standards in context we will summarize the Tuskegee Syphilis Study, a landmark case in unethical research. The Tuskegee study is important for two reasons. First, its research procedures were identified as unethical practices that regulations needed to cover. Second, it demonstrated the serious social consequences of failing to protect research subjects.

A CORNERSTONE OF ETHICAL RESEARCH: THE TUSKEGEE SYPHILIS STUDY

The Tuskegee Syphilis Study is the best known U.S. example of an egregious abuse of human subjects. The study's subjects, all of whom were poor African American men, were explicitly denied effective treatment for syphilis, that is, they were harmed simply by participating in the research.

In 1932, U.S. Public Health Service researchers started monitoring the health of two groups of African American males. One group had untreated syphilis; the other group was free of syphilis symptoms. The objective of the research was to document the course of untreated syphilis. At the time the study began, treatments for syphilis were potentially dangerous, so the researchers may have not questioned the ethics of not treating subjects. By the mid-1950s penicillin was widely available and known to be an effective treatment for syphilis. Yet the study continued until 1973, and the untreated subjects still had not received penicillin, and they had been actively discouraged from seeking treatment elsewhere. The failure to treat the subjects was particularly disturbing because the study continued unchallenged despite the Nuremberg Code.[1]

At the end of World War II, disclosure of Nazi atrocities included reports of doctors and scientists who performed human experiments on Jewish prisoners. Military tribunals were held to try the Nazi leadership. One of the Nuremberg Military Tribunal verdicts listed 10 principles of moral, ethical, and legal medical experimentation on humans. The principles, referred to as the Nuremberg Code, have formed the basis for regulations protecting human research subjects. The Tuskegee study violated the code's principles, including not asking the subjects to give their free and informed consent to participate, ignoring the researchers' obligation to avoid causing unnecessary physical suffering, not allowing the subjects to terminate their participation at any time, and disregarding the researchers' obligation to discontinue an experiment when its continuation could result in death.

Traditionally experiments include a control group, that is, a group that does not receive the experimental treatment. Ethical research practice prohibits withholding beneficial treatment from subjects. If the most beneficial treatment remains unknown, the control-group subjects must be assigned to an alternate beneficial form of treatment. Subjects cannot simply be denied treatment. For example, if you were to conduct a study of depression, all subjects would need to be assigned to some form of treatment (e.g., psychotherapy, narrative therapy, medication). Because we know so much about effective depression treatment, each form of treatment would be better than no treatment at all. If the subjects receiving an experimental treatment show marked improvements the study must be discontinued and the treatment made available to the control group. The Tuskegee study had a lasting effect in assuring that no research subject was denied beneficial treatment.

The shaded area below summarizes four other well-known studies that raised questions about whether the subjects had been treated ethically. Although no documentation exists that indicates these cases caused harm to individuals, they raised issues that have informed ethical practice.

[1]The Nuremberg Code is posted on several Internet sites. One source is http://ohsr.od.nih.gov/guide-lines/nuremberg.html. The original is from *Trials of War Criminals before the Nuremberg Military Tribunals under Control Council Law No. 10* Vol. 2 (Washington, D.C.: U.S. Government Printing Office, 1949), pp. 181–182.

Four Classic Ethics Cases Involving Human Subjects

Jewish Chronic Disease Hospital (1963).[2]

What Happened: Elderly patients were asked to give consent to be injected as part of research on immune system responses. They were not told that the injections were unrelated to their disease or its treatment or that the injections contained live cancer cells.

Ethical Concern: Investigators found that vague request for consent did not constitute informed consent.

University of Chicago Law School Taping of Jury Deliberation (1954)[3]

What Happened: Discussions among juries hearing civil cases were recorded without their knowledge, but with the permission of the judge and litigant's attorneys. Recordings were to be kept with judge until the case was closed.

Ethical Concern: Potential loss of confidence in public institutions by compromising the secrecy of jury deliberations.

The Tearoom Trade (1970)[4]

What Happened: A doctoral student served as a lookout and observer for men having sexual encounters in public restrooms. Later he used license plate numbers to track down the men and interview them under the pretense of conducting a public health survey.

Ethical Concern: An example of deceptive research, where a subject is not told the purpose of the research and potentially the researcher discloses something the subject considers private.

The Milgram Experiments (1961)[5]

What Happened: Subjects were recruited to give other subjects (actually actors) what they thought were electric shocks.

Ethical Concern: Another example of deceptive research. In this case subjects might gain unwanted self-knowledge. ■

PRINCIPLES OF ETHICAL TREATMENT OF HUMAN SUBJECTS

In response to the Tuskegee study and other reported abuses, the U.S. government formed the National Commission for the Protection of Human Subjects of Biomedical and Behavioral Research and charged it with identifying the basic ethical principles for research involving human subjects. The commission summarized its findings in the Belmont Report.[6] The report identified three basic ethical principles: respect for persons, beneficence, and justice. Respect for persons requires that subjects participate voluntarily and with adequate

---

[2]J. Katz. *Experimentation with Human Beings* (New York: Russell Sage Foundations, 1972) pp. 10–65.

[3]*Ibid.*, pp. 68–103.

[4]*Ibid.*, pp. 325–329.

[5]*Ibid.*, pp. 358–365.

[6]"The Belmont Report Ethical Principles and Guidelines for the protection of human subjects of research," The National Commission for the Protection of Human Subjects of Biomedical and Behavioral Research, U.S. Department of Health, Education, and Welfare, April 18, 1979 (available at http://ohsr.od.nih.gov/guidelines/belmont.html, accessed June 12, 2008).

information about the proposed study to give informed decisions regarding participation. Beneficence requires doing no harm and maximizing possible benefits and minimizing possible harms. Justice requires that research subjects not be selected "simply because of their easy availability, their compromised position, or their manipulability, rather than for reasons directly related to the problem at hand."

These principles underlie U.S. regulations, which apply to investigators conducting research for a federal agency or on a federally funded project. The principles should guide the behavior of all members of a study team whether or not the study is covered by federal regulations. Implementation of these principles requires that subjects give informed consent, that benefits and risks be identified and weighed, and that selection of subjects be fair. Informed consent demonstrates respect for persons; assessing risks and benefits demonstrates beneficence, and fairly selecting subjects demonstrates justice.

Informed Consent

A potential research subject must have adequate information to make an informed, voluntary decision to participate. You must tell potential subjects in words that they can clearly understand about the study's purpose, its potential risks, and possible benefits. They need to know what they will be expected to do, what will be done to them, and what will be done with the collected information. You should provide them with information that answers the following questions. Who will receive the information? What type of information will be disseminated? What steps will be taken to protect the identity of the subject? What will happen if you learn of an illegal act or identify a health risk? If photos, movies, recordings, or similar research records are being produced, what will be done with them? We know of one student who was shocked to learn that an interview tape, produced as part of an experiment, was being shown to classes at her college.

Specific circumstances may impede a subject's ability to make a voluntary decision. Prisoners, members of the military, and schoolchildren, all of whom are in controlled settings, may interpret requests for information or participation as commands. Research involving prisoners is particularly difficult. Even modest incentives can compromise a prisoner's ability to assess the risks of participation and may preclude a voluntary decision to participate. Similarly, patients may mistakenly believe that their research participation will have a therapeutic benefit. They should receive clear, realistic information on the benefits and risks of participation. Still, the ability of seriously ill persons to give informed consent, no matter what they are told, is questionable. A *New York Times* article summed up the problem as follows: "Potential participants are often desperately ill and may grasp at any straw—even signing a document without reading it. For this reason, many say there is no such thing as informed consent—only consent."[7] Vulnerable populations, such as children, aged people, and mentally disabled people, may not be

---

[7]L. K. Altman, "Fatal Drug Trial Raises Questions about 'Informed Consent,'" *New York Times*, October 4, 1993, B7.

fully capable of making an informed decision or protecting their own interests. In general, researchers try to get informed consent from both the potential subjects and from their legal guardians.

An authority relationship between the investigator and potential subjects may raise questions about voluntary consent. Physicians, social workers, or professors may ask their patients, clients, or students to participate in a research study. A patient may agree because she doesn't want to sour the relationship with her physician. A client may agree because he may suspect that he will be overlooked for other opportunities to alleviate his problems. A student may agree because she suspects that it won't hurt and may help her course grade.

Finally, voluntary participation requires the ability to withdraw from a study at any time. Potential subjects must be told this as part of the process of obtaining informed consent. Furthermore, they must also be told that other benefits they are entitled to will not be affected by their decision to participate or to discontinue their participation. For example, a client receiving social services must be told that his continued eligibility for these services does not depend on participating in a research study.

Informed consent and a signed informed consent form are not one and the same. A subject must be free to make her own decision about whether she wishes to participate or not. To make this decision she must be informed about the study and give her consent to participate. A signed consent form merely documents that she received the information and consented to participate. For projects where a subject experiences no risks beyond the risks of everyday life or ordinary professional responsibilities, signed statements may be reasonably viewed as unnecessary. Online surveys typically open with a statement describing the research; subjects indicate their willingness to participate by clicking on an "accept" button. Below we give you an example of obtaining consent for an opinion survey. Just because the recipient of a survey does not sign an informed consent form does not relieve you of the responsibility of giving the subject sufficient information so her consent to participate is truly informed.

Identifying and Weighing Costs, Risks, and Benefits

People may not agree on the risks and benefits of research participation. Individuals' educational, social, and professional backgrounds contribute to how they define and rank risks and benefits. You may incorrectly assume that a proposed study presents minimal or no risk. You should be especially vigilant if potential

---

Request for Consent to Participate on an Online Survey

This is a study about opinions regarding various issues in the news. This is an anonymous survey and your identity is not connected to your responses in any way. Clicking "Yes" below indicates that you agree to participate in this study.

☐ Yes, I agree to participate.
☐ No, I do not wish to participate. ∎

subjects differ from you. For example, you may underestimate the potential harm in studying recent immigrants. You may misjudge what constitutes a risk or a benefit for an immigrant and erroneously assume that your requests for consent are unbiased and informative.

What are the benefits of participating in research? The research may involve a treatment or program expected to relieve a psychological problem or increase economic opportunities. The research may seem to benefit a group that a potential subject values. Survivors of floods and wildfires may agree to participate in studies because they believe that the findings will contribute to better future response and recovery efforts. For some studies—perhaps most—the subject may participate because the research question seems somewhat interesting and the inconvenience is minimal.

Remuneration may be a valuable benefit. We know of a few graduate students who subsidized their incomes by participating as subjects for biomedical research studies. Paying subjects for the inconvenience of participating is not unethical, unless the amount is so large that it may be viewed as a bribe or a questionable inducement to participate. What distinguishes reasonable reimbursement from "questionable inducements"? Federal regulations offer no guidance, and opinions vary as to whether participants should receive more than compensation for their direct costs.

What are the risks of participating in research? The most commonly cited risks are physical harm, pain or discomfort, embarrassment, loss of privacy, loss of time, and inconvenience. A study could potentially uncover illegal behavior. As part of informed consent potential subjects must be told what risks they may experience during the study or as a result of the study. They must be told if the risks are unknown, or if researchers disagree on the risks.

Subjects can only be told about benefits that can be reasonably expected. Theoretically, a study may be groundbreaking; however, most studies are not. Consequently, a subject should not be told that a study has a probability of generating significant knowledge. Nor should potential subjects be led to believe that they will gain benefits that are possible but unlikely.

Although not usually covered as part of informed consent you may wish to consider that risks and benefits may apply beyond individuals. For example, families may be affected by the time a family member spends participating in a study, by his reactions after sharing personal information, or by the outcome of participating in a program to encourage lifestyle changes.

Selection of Subjects

Recruiting subjects is the first step in assuring informed consent. Fliers, Internet sites, or other recruiting materials and media should state that participants are sought for a research project. The words used to solicit subjects may act as a questionable inducement. Consider the attractiveness of being asked to test out a "new" or "exciting" treatment, to participate in a "free" program, or to receive "$1000 for a weekend stay in our research facility." If you recruit participants through personal contact, you must be particularly sensitive to not pressuring them to participate.

Simply contacting potential subjects can raise privacy concerns. *Privacy* refers to the ability to control disclosure and dissemination of information about oneself. One dimension of privacy is physical privacy, that is, not having to endure unwanted intrusions. If a person is likely to wonder "how did they get my name?" you may have violated his privacy. For example, if you plan to collect information from food pantry users, a staff member who knows the users should ask their permission to be contacted. They should be assured that whether or not they choose to participate will not affect services they normally receive.

Sample size is not included as a component of informed consent, nor is it discussed in behavioral and social science texts on research ethics. Still, the desired number of subjects may affect how vigorously subjects are recruited and open the door for subtle coercion. The American Statistical Association includes sample size in its ethical guidelines. Its guidelines state that statisticians making informed recommendations should avoid allowing an excessive number of subjects or an inadequate sample.[8] An inadequate sample can affect the quality of the statistical analysis, for example, decisions about whether findings occurred by chance. Conversely, an overly large sample may squander resources, including participants' time. An appropriate sample size partially depends on how the data will be analyzed. For example, a sample of 400 might be adequate to describe volunteer activities of state residents, but it may be too small if you want to compare volunteers in the state's counties.

PROTECTING PRIVACY AND CONFIDENTIALITY

You may encounter the terms *privacy, confidentiality, anonymity,* and *research records* in discussions on collecting individual data. As we noted, *privacy* refers to an individual's ability to control the access of other people to information about himself. *Confidentiality* refers to protection of information, so that you cannot or will not disclose records with individual identifiers. *Anonymity* refers to collecting information so that you cannot link any piece of data to a specific, named individual. *Research records* refer to records gathered and maintained for the purpose of describing or making generalizations about groups of persons. Research records are different from administrative or clinical records, which are meant to make judgments about an individual or to support decisions that directly and personally affect an individual.

The requirements of voluntary participation and informed consent uphold the individual's right to have control over personal information. While you may promise anonymity or confidentiality, a potential subject may not necessarily trust you to follow through. Guarantees of confidentiality neither ensure candor nor increase propensity to participate in research; rather, limited research has found that respondents tend to view promises of confidentiality skeptically.[9] Nevertheless, you must

---

[8]American Statistical Association. *Ethical Guidelines for Statistical Practice* (1999). The guidelines are available on the association's Web site (www.amerstat.org/profession/ethicalstatistics.html).

[9]A. G. Turner, "What Subjects of Survey Research Believe about Confidentiality," in *The Ethics of Social Research: Surveys and Experiments,* J. E. Sieber, ed. (New York: Springer-Verlag, 1982), pp. 151–165. For a summary of more recent research on the public's attitude about the confidentiality of data see *Expanding Access to Research Data: Reconciling Risks and Opportunities* (Washington, D.C.: National Academies Press, 2005) pp. 52–54.

respect participants' privacy and maintain confidences as part of your professional responsibilities to subjects.

Some questions may seem unduly intrusive. They may stir up unpleasant recollections or painful feelings for subjects. Such topics include research on sexual behaviors, victimization, or discrimination. People who read pornography, have poor reasoning skills, or harbor controversial opinions may prefer to keep this information to themselves. Disclosure of behaviors such as drug use, child abuse, or criminal activity may cause a respondent to fear that she will be "found out." For a study of a sensitive topic to be ethical (1) the psychological and social risks must have been identified, (2) the benefits of answering the research question must offset the potential risks, (3) the prospective subjects must be informed of the risks, and (4) promises of confidentiality must be maintained.[10]

In deciding whether to keep responses confidential or to collect anonymous data you may first consider subject anonymity, that is, ensuring that no records are kept on the identity of subjects, and data cannot be traced back to a specific individual. If you are conducting a study for an agency, an approach that approximates subject anonymity is for agency staff to select the sample and distribute questionnaires or collect information from agency files. The staff should delete any identifying information from the records before allowing you to examine the records.

Often, however, anonymity is impossible. You must know subjects' names to follow up on nonrespondents or to compare respondents and nonrespondents; to combine information provided by a subject with information from agency records; or to collect information from an individual at different points in time. Subject names must be available for research supervisors or auditors to verify that the research was done and that accurate information was collected and reported. Although research auditing can raise concerns about confidentiality, without audits, incompetence or malfeasance may go undetected.

Confidentiality may also be breeched by carelessness, legal demands, or statistical disclosure. To avoid accidental disclosure of personal information, identifying information should be separated from an individual's data. A code or alias can be assigned to each record and the list of names and codes stored separately in a secure place. For longitudinal studies and other studies that collect information from more than one source, aliases can be used to identify individuals and combine the individual information. A respondent may choose her own alias, that is, information that others cannot easily obtain, such as her mother's birth date. The success of having respondents choose their alias depends on their ability to remember each time they are asked.

Theoretically, researchers can have their records subpoenaed. Federal policies offer some protections to participants and researchers. The Confidential Information Protection and Statistical Efficiency Act (CIPSEA) requires that federal statistical agencies inform respondents and get their informed consent if the information

---

[10]For an extensive discussion of procedures to protect privacy, applicable federal laws, and an extensive bibliography, see *Protecting Human Research Subjects*, pp. 3-27–3-37, 3-56. Readers interested in strategies for identifying and questioning subjects about sensitive topics may wish to read the cases in C. M. Renzetti and R. M. Lee, eds., *Researching Sensitive Topics* (Newbury Park: Sage Publications, 1992).

is to be used for nonstatistical purposes.[11] Certificates of confidentiality by and large protect researchers investigating sensitive subjects, such as mental illness or drug abuse, from having to provide identifying information. The U.S. Department of Health and Human Services offers certificates to researchers whose subjects might be harmed if they were identified. The Department of Justice offers certificates to researchers conducting criminal justice research. However, a certificate cannot be obtained after data collection is completed.[12]

When reporting data or sharing research records, you should be sensitive to unintended disclosures. For example, when reporting on a case study you should consider using pseudonyms and altering some personal information, such as occupation. Before any research records leave your control, you should identify potential breaches of informed consent and confidentiality. A subject may have agreed to participate for a specific purpose, without giving blanket authorization for other uses. As part of obtaining informed consent you should identify anticipated future uses of the data, including their availability for independent verification of the study's implementation and replication of its analysis.[13] Before releasing data, make sure that identifiers such as names, addresses, and telephone numbers have been removed.

A more formidable problem is that of *deductive disclosure.* If the names of participants in a study are known, someone could sort through the data to identify a specific person. With a list of respondents, someone may sort the data by age, race, sex, and position and deduce a respondent's identity. One way to protect against such abuses is to not disclose the list of respondents. Deductive disclosure is a concern with publicly available electronic databases. For example, if census data were released as reported, one could learn detailed information about the only three Hispanic families in a county or a state's six female-owned utilities companies. To prevent such abuses the Census Bureau has developed procedures to release as much information as possible without violating respondents' privacy. As is true with many of the Census Bureau's statistical practices these procedures can serve as a model.

An emerging concern is the ethical implications of Internet research. Research might be done on Web sites, electronic bulletin boards, Listservs, or social networking sites. It may be naïve to assume that any communications sent out into cyberspace is private, and there is no consensus about what online communications should be considered public as opposed to private. Information found on a Web site may be treated the same as other textual material, that is, you would not normally seek informed consent or be concerned about privacy.

---

[11]For a brief, but fuller, discussion, of federal legal protections, see *Expanding Access to Research Data*, pp. 56–59. The report also considers threats to confidentiality associated with national security concerns.

[12]*Ibid.*, p.56

[13]T. E. Hedrick, "Justifications and Obstacles to Data Sharing," in S. E. Fienberg, M. E. Martin, and M. L. Straf, eds., *Sharing Research Data* (Washington, D.C.: National Academy Press, 1985), p. 136. Hedrick cites sources that discuss this issue in more depth. See also the *Ethical Guidelines for Statistical Practice*, D4.

FEDERAL POLICY ON PROTECTION OF HUMAN SUBJECTS AND INSTITUTIONAL REVIEW BOARDS

In 1991 a uniform federal policy for the protection of human subjects, the Common Rule, was published. Its hallmark was a requirement that every institution receiving federal money for research involving human subjects create an *Institutional Review Board* (IRB) and appoint its members. The IRB determines if a proposed project meets the following requirements: (1) risks to subjects are minimized, (2) risks are reasonable in relation to anticipated benefits, (3) selection of subjects is equitable, (4) informed consent will be sought and appropriately documented, (5) appropriate data will be monitored to ensure the safety of subjects, and (6) adequate provisions exist for ensuring privacy of subjects and confidentiality of data.[14] Two of these criteria merit further mention. First, the long-range effects of the knowledge gained from the research are explicitly excluded in determining the risks and benefits of participation. Second, the need to address possible abuses of vulnerable populations is stressed. IRBs are encouraged to consider whether research on a specific population is consistent with an equitable selection of subjects and to make sure that these populations' vulnerability "to coercion or undue influence" has not been exploited.

An IRB reviews all research under the purview of the institution that involves human subjects. To review only publicly or privately supported research would imply that only funded projects have to conform to ethical practices. An institution that fails to comply with the Common Rule can have its federal funding terminated or suspended. Some university IRBs have been accused of being overly cautious and throwing unnecessary roadblocks in the way of research that includes surveys or field studies, to avoid potentially harmful consequences.[15]

To what extent do you have to concern yourself with IRB review? Federal policies protecting human subjects require compliance by federal agencies, institutions, and individual researchers. If you are a student or an employee of a university, a medical facility, or other institution that receives federal research funds and you will be conducting a study that involves interacting with people, asking them to do something, or using identifiable private information, consult with your IRB before you start your research. Putting off learning about your institution's IRB procedures can lead to long delays and frustration while conducting your study.

The IRB chair or the chair's designee will determine if your project is exempt from further review, appropriate for an expedited review, or must be subject to full review. Projects exempt from review or eligible for an expedited review receive less close scrutiny. The categories of exempt projects or expedited review allow research involving minimal risk to proceed and avoid the long delays associated

---

[14]45 CFR 46 Section 46.111.

[15]For a discussion of the effect of IRBs on social science research see R. J-P. Hauck, ed., "Symposium: Protecting Human Research Participants, IRBs, and Political Science Redux," *Political Science and Politics*, 41 (July 2008): 475–511.

with a full IRB review. *Minimal risk* applies to those projects where the risks of participating in the research are similar to the risks encountered in daily life. For example, an IRB may waive written documentation of informed consent for surveys and observation of public behaviors. Waivers, however, are not permitted if responses could be traced to a specific individual and if disclosure could result in civil or criminal liability or damage subjects' financial standing, employability, or reputation. The federal regulations are more detailed than what has been presented here. Furthermore, the regulations receive frequent scrutiny, and the practices are still evolving. Members of an IRB or the National Institutes of Health Office of Human Subjects Research Web site (http://ohsr.od.nih.gov) are good sources for detailed or up-to-date information.

BEYOND INFORMED CONSENT AND CONFIDENTIALITY: ISSUES OF INTEREST TO ADMINISTRATORS

Public agencies and nonprofits that do not conduct federally funded research may not have an IRB. Still, the agencies should adhere to the values and practices identified by the Belmont Report. Recall that these values and practices were

Respect for persons requires informed consent;

Beneficence requires that benefits outweigh risks;

Justice requires a fair selection of subjects.

These values cut across professions, disciplines, and organizations. Whether you are an investigator or part of management you should ascertain that studies intended to describe groups of persons adhere to these principles and practices. The practices specifically apply to research efforts; they do not necessarily apply if the data are collected for administrative or clinical records. For example, ethical research practices do not require supervisors get informed consent from an employee to evaluate her or to demonstrate that the benefit of preparing and conducting a performance evaluation justifies its cost.

You and others involved in deciding whether to implement a study may ask the same questions that are asked to obtain informed consent. There are no right or wrong answers, but the answers will help you decide if the informed consent content and procedures are adequate, if the risk of participation is outweighed by the benefits, and if selection of participants is fair and unbiased. Key questions include the following: Will subjects be anonymous or confidential? What steps will be taken to protect their identities? What will they be asked to do? How much time will it take? Are there other potential risks? How are subjects going to be recruited and selected? What information will potential subjects be given to obtain informed consent? How will informed consent be obtained?

You should consider how the proposed research may affect the agency's reputation. No matter what the agency's role, its reputation may be enhanced by research that others consider valuable and harmed by research that others consider worthless, intrusive, or harmful. You may question whether a planned study will unduly infringe on respondents' privacy or abuse their time. You should decide if a study requiring agency resources, including time, represents a good use of money or donor contributions. You should be convinced that a study will likely yield valued information. You should decide if assumptions that others will act on the findings are realistic.

Unreliable items, unwanted items, or unused studies waste respondents' time. Items not operationally valid may abuse respondents' goodwill, insofar as their responses contribute to incorrect or misleading conclusions. The detrimental effects of an unwanted or poorly designed study go beyond the respondents. Future studies also are affected if potential subjects become cynical about research and the value of their participation.

Seeking too much information or seeking it too often can build resistance to future requests for information. Consider the complaints of businesses and state and local governments that churn out data only to meet federal information requirements or the frustration of nonprofits that have to produce frequent reports to assure funders that their money is being well spent. If individual respondents perceive that the data are merely collected but not used, they may not only complain but also refuse requests for information.

The agency is responsible for protecting the privacy of its employees or clients. Potential subjects should be contacted by an agency representative, possibly by some-one from the unit sponsoring the study. The agency representative should explain the nature of the study, get permission to give the potential subjects' contact information to investigators, and in noncoercive language indicate the value of the subjects' responses. Clients should be clearly told that if they decline to participate they will not lose eligibility for agency services. Public announcements, such as posters, may be used to recruit employees or clients. Voluntary participation is less likely to be compromised with posted announcements than by personal solicitation.[16]

For employee participation to be voluntary, whether or not employees decide to participate should not affect performance ratings, pay, or similar decisions. This should be clearly communicated to the employee.[17] You should not assume that assurances will overcome employee beliefs that refusal to participate will have con-sequences or that promises of confidentiality will not be kept. If an employee plans to conduct research as part of a graduate program, the research must be vetted by the university's IRB. The agency administrators should also consider how an employee's research could impact the organization and participating employees. Is the research taking time away from other tasks? How will participants interpret the topic? Do they assume that the organization is going to address a long-standing problem? Or are they anxious that a major change is in the works?

Demonstration projects bring up concerns about what will happen once the project ends. Projects that are part of a research grant may be funded for a specific period of time, but the participants' need for services may continue. In prisons or psychiatric hospitals, participants in an experimental program may feel abandoned if the program ends once the data are collected or the funding stops. A related issue is how participation in a demonstration project will affect the clients. For example, clients enrolled in a demonstration project that provides job training may find that

---

[16]*Protecting Human Research Subjects*, p. 6-53.
[17]*Ibid.*, p. 6-55.

no employer can use their new skills or that the training has not qualified them for advanced courses.

Ideally, relevant stakeholders should review a proposal. Input from representatives of the participant community are especially valuable in identifying potential risks. For example, parolees, ex-convicts, and prison guards might review proposals for studies involving prisoners. Even apparently innocuous groups such as employees may identify unexpected concerns. Stakeholders may be helpful in deciding if potential subjects will understand the purpose of the study and what is being asked of them and the risks involved.

You should also check that subjects will be debriefed, if appropriate, at the end of their participation. For example, subjects who are asked to perform job-related tasks may be disappointed or frustrated about their performance. A debriefing offers the researcher an opportunity to observe any negative effects of the research and to answer questions or concerns that a subject may have.

Another issue is what will become of the research documents. What will be done with the completed questionnaires, recordings, or other research materials?[18] Will they be put into an archive? If so, the subjects should be told where the data will be stored and how individual identities will be protected. If the data are being collected by a consultant, will the agency receive completed questionnaires, spreadsheets, or electronic databases? Will the consultant remove individual identifiers? Potential subjects need to know who will have access to their information before they can give informed consent.

ETHICAL VALUES[19]

Ethical values affect us when we gather information and when we report our findings. The ethical issues associated with reporting are covered in Chapter 14. The following cover the major values that should guide our behavior as researchers as we prepare to conduct research.

Honesty: Strive for truthfulness in all scientific communications. Honestly report data, results, methods and procedures, and publication status.

Integrity: Fulfill promises and agreements. Act in good faith.

Confidentiality: Protect the communications of participants. Do not disclose personnel records, trade or military secrets, and patient records.

Nondiscrimination: Do not provide preferential treatment to participants on the basis of sex, race, ethnicity, or other factors that are not related to their scientific competence and integrity.

Competence: Understand and do not exceed your own professional capacity and limits.

---

[18]Paul Oliver, *The Student's Guide to Research Ethics* (Philadelphia: Open University Press, 2003). Chapter 4 has a useful discussion related to record storage concerns.
[19]Adapted from A. Shamoo and D. Resnik. *Responsible Conduct of Research* (New York: Oxford University Press, 2003). *Source*: http://www.niehs.nih.gov/research/resources/bioethics/whatis.cfm

Legality: Know and obey relevant laws and institutional and governmental policies.

Human subjects protection: When conducting research on human subjects, minimize harms and risks and maximize benefits; respect human dignity, privacy, and autonomy; take special precautions with vulnerable populations; and strive to distribute the benefits and burdens of research fairly.

CONCLUDING OBSERVATIONS

Recognizing and protecting research subjects is imperative whether you are an investigator or a manager. In the event that federally funded research is being conducted, an IRB review is required. This process helps to make sure that you have appropriately identified risks and benefits, communicated them to potential subjects, and have adequate procedures to obtain informed consent. Although you may not have an institutional review board to guide your research, you must follow ethical practices in administering surveys, conducting interviews, or collecting data from records.

We may erroneously assume that research that does not cause physical harm and does not have a negative effect. Wasting subjects' time is harmful, so is invading subjects' privacy. Respect for persons, beneficence, and justice may seem like lofty ideals, but in practice as you apply these concepts you may identify and address potential problems. You will also save your agency from possibly wasting resources, diminishing its reputation, and facing litigation. The bottom line is that you should protect research subjects because it is the right thing to do.

RECOMMENDED RESOURCES

Paul Oliver, *The Student's Guide to Research Ethics* (Philadelphia: McGraw Hill, 2003).

Jay Katz, *Experimentation with Human Beings* (New York: Russell Sage Foundation, 1972). Although 40 years old this is an excellent resource on cases that informed discussions about protecting human subjects.

James F. Childress, Eric M. Meslin, and Harold T. Shapiro, eds. *Belmont Revisited: Ethical Principles for Research with Human Subjects* (Washington,

D.C.: Georgetown University Press, 2008). This resource carries a series of essays that cover the development of the nearly 40-year-old Belmont Report and its impact on current discussions on protecting human subjects.

To keep up to date with federal regulations, visit the Web site of the Office of Human Research Protections within the U.S. Department of Health and Human Services: http://www.hhs.gov/ohrp/

CHAPTER 3   EXERCISES

This chapter has three exercises that ask to you consider ethical research practices in different contexts.

- Exercise 3.1 Learning about Teenaged Mothers asks you to identify and address ethical concerns in conducting a study with adolescent participants. You are also asked to review and comment on how the researchers might solicit participants' informed consent.
- Exercise 3.2 Evaluating a Debt Counseling Program asks you to consider ethical concerns in a study that involves both staff and client participants.

- Exercise 3.3 On Your Own asks you to identify your agency's protocol or practices about conducting research. This exercise is applicable if you are planning to do a study as part of your job or internship and has questions identifying factors that are part of the process of assuring that human subjects are protected.

EXERCISE 3.1 Learning about Teenaged Mothers

Scenario

Elaine Bell Kaplan, the author of *Not Our Kind of Girl* (Berkley, CA: University of California Press, 1997), conducted a study of African American teenage mothers. Her purpose was to examine the causes and consequences of Black teenage parenthood. She gave 32 teenaged mothers living in two California communities a 126-item questionnaire that asked about their experiences before, during, and after the birth of their children. Twenty-five participants were interviewed. Interviews included questions about their family, sexual behavior, the child's father, their experiences with school and welfare agencies, and quality of life. The interviews were audio-taped and transcribed by the researcher.

Section A: Getting Started

1. Based on this information what ethical considerations should be most dominant for the researcher?
2. The following questions explore how you would ensure that you have treated your subjects ethically.
   a. What kind of procedures would be necessary to make sure that the participants in the study are not harmed?
   b. What types of harm might you anticipate? What protections would you build into the research to address such possible ill-effects?
3. How would you protect access to the participants' audio-taped interviews?
4. If you were doing a similar study of adolescent mothers what ethical considerations would affect how you go about recruiting subjects?
5. What ethical considerations would guide your decisions about offering the subjects incentives? What kind of incentives would be reasonable to offer to this population?
6. What follows is a hypothetical informed consent form for a similar study being conducted by a county agency. Assume that you are Alpha Greene's supervisor.
   a. Review each section of the form and assess its strengths and weaknesses.
   b. Identify and discuss the changes you would recommend.

Informed Consent Form for Research

Title

Experiences of Teenage Mothers in Aries County, California

Principal Investigator

Alpha Greene, Research Analyst, Department of Human Services, Aries County, CA

We are asking you to participate in a research study. The purpose of this study is to learn from teen mothers in Aries City about their pregnancy and the challenges of raising a baby. We would like to explore with you (1) what your life was like before you became pregnant, (2) the challenges you faced during pregnancy and how you handled them, (3) how life has been going for you and your child since you became a mother, (4) your relationship with the baby's father, and (5) your experiences with your school and community human service agencies. The study will compare the answers of teenage methods from different population groups, for example, African American mothers and Hispanic mothers.

Information

If you agree to participate in this study, you will be asked to answer a 126-item questionnaire and participate in an in-person interview that will be tape recorded to assure the accuracy of the information. The questionnaire should take 30 minutes to finish; the interview should last no longer than 90 minutes.

Risks

There are no foreseeable risks to participating in this research.

Benefits

There is no direct benefit to you for participating. Your responses will help the county plan and implement effective programs in schools and human service agencies.

Confidentiality

The information in the study records will be kept strictly confidential. Data will be stored securely in a locked file cabinet in the principal investigator's office. No reference will be made in oral or written reports that could link you to the study.

Compensation

There is no compensation for participating in this study.

Contact

If you have questions at any time about the study or the procedures, you may contact the researcher [contact information given]. If you feel you have not been treated according to the descriptions in this form, or your rights as a participant in research have been violated during the course of this project, you may contact the Department of Human Services Associate Director [contact information given].

Participation

Your participation in this study is voluntary; you may decline to participate without penalty. If you decide to participate, you may withdraw from the study at any time without penalty and without loss of benefits to which you are otherwise entitled. If you withdraw from the study before data collection is completed your data will be returned to you or destroyed at your request.

Consent

"I have read and understand the above information. I have received a copy of this form. I agree to participate in this study with the understanding that I may withdraw at any time."

Participant's signature_____   Investigator's signature _____

Date_____   Date _____

Section B: Class Discussion

As part of preparing for the class discussion, learn if your university has an IRB and if so what its requirements are. For Exercises 1 and 2 you may want to consider in your discussion whether you would handle informed consent differently to meet university requirements instead of conducting the study for an agency.

1. In small groups, review the informed consent form. Each group should present a strategy to the class for obtaining informed consent from the study participants. The strategies should include (a) the wording of the form and (b) how the form should be presented to potential subjects.

EXERCISE 3.2 Evaluating a Debt Counseling Program

Scenario

You have been asked to evaluate an organization's debt counseling program. Several different studies may be undertaken, including the following: an assessment of staff competence, for example, the quality of advice offered, appropriate follow-up, ability to work with diverse clients, whether clients benefit from the program, and what separates clients who benefit and those who don't.

Section A: Getting Started

1. Consider how the ethical principles discussed in this chapter apply to staff and client participants.
2. What type of harm might staff participants experience because of your study?
3. What types of harm might client participants experience because of your study?
4. What protections would you build into the research to address possible ill-effects?

Section B: Small Group Exercises

1. As a class or in small groups, develop a research plan and evaluate how well it protects staff and client participants. The plan should include recruitment of subjects, obtaining informed consent, maintaining assurances of confidentiality or anonymity of information, analysis of information, and storage of data.
2. Based on your research plan and method of obtaining informed consent how would you handle the following situations?
   a. As part of collecting the information you learn that a specific staff member is consistently giving erroneous advice. Does it matter if the staff member was a research subject or not? Justify your opinion.
   b. As part of collecting the information you learn that a client who is a research subject has lied about his financial status.
   c. After your report is given to the organization, evidence surfaces that it has not taken action on the report's findings about serious problems that affect program quality.

EXERCISE 3.3   On Your Own

If you are planning to conduct a study for your agency you first should find out what kind of ethical protocols it has. If your agency does not have an IRB, more than likely the study will have to be reviewed by administrators who would certainly have lots of questions about your research. The answers to these questions will help you consider how to proceed in an ethical manner and protect your human subjects.

1. What methods will you use to collect data (e.g., what will participants be asked to do)?
2. How much time will it take?
3. Will this study harm the reputation of the organization if the responses are not as favorable as we would like?
4. Will the methods you use allow for anonymity or confidentiality?
5. How will you explain the caveats to confidentiality for your participants (e.g., duty to warn)?
6. How will participants' identities be protected?
7. If this is a demonstration project, how will you make sure that the selection process is fair and equitable?
8. What other questions do you anticipate from your administrator?

Performance Measurement Systems: Their Design and Analysis

In Chapters 4 and 5 we apply research methodology to a common management tool—measuring performance. Measuring people's and organization's performance is so common that you may not realize how much it depends on methodological decisions. In Chapter 4 we embellish our discussion of measurement to discuss how you can identify appropriate measures and find data sources. We also introduce time series, a research design that guides how you present and interpret performance data.

Chapter 5 covers the graphs and statistics that you can use to analyze and present performance measures. This chapter will serve as a review insofar as you encountered most of these graphs and statistics in your youth. More importantly, this chapter demonstrates there is not just one way or even "one best way" to present performance data. The choice of what graph or statistic best describes the data is yours. In addition to analyzing performance data you can use these graphs and statistics in analyzing data for other management decisions.

Designing Performance Measures and Monitoring Systems

Y ou may have heard the phrases "work smarter not harder" or "results-based management." How do managers know how hard or how smart a person or group is working or whether a program is achieving results? They begin by collecting data that measure work and accomplishment. Collecting data on what a work unit, program, or organization is doing is called *performance measurement* or *performance monitoring*. Let's assume that you want to learn more about the performance of a food pantry. (Food pantries distribute food to families and individuals at risk of hunger.) You might want to know how many pounds of fruits and vegetables does it have available for distribution? How many households does it serve? How many food pantry users eat a nutritious diet?

To answer these questions you need to apply several research skills. First, you can build on what you have learned about measurement in Chapter 2. You need reliable, operationally valid, and sensitive measures to gather data on the amount of food, the number of households, and the incidence of nutritious diets. You have to decide how often to collect and compile the data. Then you have to analyze and interpret what you find.

In this chapter we present skills that link research methodology to information that managers use daily. Specifically, we consider research tools to create and implement programs and monitor their performance by focusing on three topics that apply to performance monitoring and other quantitative studies. First, we present logic models, which play a role in designing, monitoring, and evaluating programs. The models link a program's activities to its achievements and help you select operationally valid measures to track them. Second, we identify data sources commonly used to measure performance. Third, we discuss time series, a research design to present and interpret data over regular intervals. This chapter, however, does not provide a comprehensive introduction to performance measurement and monitoring systems; instead it lists resources that cover these topics in depth, in the Recommended Resources at the end of the chapter.

AN OVERVIEW OF PERFORMANCE MEASUREMENT AND MONITORING

A performance-monitoring system requires data. In deciding what data to include, remember that just collecting data may not provide useful information. Good managers want information that enables them to understand how their programs work, to communicate this to others, and to take corrective action if necessary. To create useful information you must (1) decide what to measure and when and how to measure it and (2) select statistical tools that effectively present the data and facilitate comparisons over time. As a strategic manager you can use performance information to help assign staff, develop an annual budget, or assess the effectiveness of a program. You can tell funders how their money was spent and what good it did. You may rely on the data to create an annual report, which shows what was done with what resources and for what recipients, and how these figures compare with those of previous years.

You may produce and monitor performance data to identify how often services are provided, the number of people served, the cost of providing the services, and the benefits generated. You may want to review information at regular intervals so you can monitor changes in how an organization or program uses its resources, responds to demands for service, and makes progress toward achieving its goals. For example, the manager of a food pantry may want to know: how the number of clients and their characteristics vary from one month to the next over the course of a year; if the availability of nutritious food has improved the diet of pantry users; or if the number of food-insecure families has changed over time. He may also want to know the average cost of providing one family with food and how this cost has changed over time.

LOGIC MODELS

All too often, programs are developed because someone has what sounded like a good idea. One such good idea, having children wear school uniforms, was advocated to decrease troublesome behavior. However, school uniforms have little or no effect on children's behavior.[1] Logic models may decrease the number of similar "good" ideas that waste public monies. As you build a model you and others should answer the question "why should this idea (a program or strategy) have a positive impact on a particular social condition?" Existing research may provide the answer, but often you have to rely on logic and thoughtful debate. Relying on existing research, logic, and debate, a logic model requires participants to explain why specific activities are likely to produce a particular effect and why certain resources are needed. A logic model may help you propose effective programs, formulate a strategy to implementation them, know what resources you need to amass and deploy, and select performance measures to provide feedback and

---

[1]E. Gentile and S. A. Imberman. "Dressed for success: Do school uniforms improve student behavior, attendance, and achievement?" University of Houston Working Paper 2009-03, posted at *www.uh .edu/econpapers/RePEc/hou/wpaper/2009-03.pdf*

demonstrate accountability. In our own experience logic models are useful from the time an organization plans a program through to its evaluation. Logic models are valuable in deciding what an organization should measure and monitor.

Simply stated, a logic model links the components consisting of a program's resources, activities, and service units to an expected outcome. To build a logic model you should work with a management team and other relevant stakeholders. Start with the organization's mission statement and identify how the program will help the organization achieve its mission. Typically, you should articulate what type of change you expect a program to produce. It may be a change in program participants, agency operations, or the community. You next identify the activities that must take place for the anticipated change to occur. You then identify the resources needed to conduct the activities as planned. Depending on why the model is being built you may set annual goals and estimate how long it will take to reach the long-term outcome. Figure 4.1 presents a general outline for a logic model and labels its components.

The logic model components are identified by conventional terms used to describe programs. Understanding these terms may help you to establish the operational validity of your measures. (Recall from Chapter 2 that operational validity requires that you ask if you are measuring what you intended to measure.) The concept of operational validity is important as you think about what constitutes an activity, an output, or an outcome. The common definitions of the related terms are

> *Inputs*—resources needed to operate the program. These include staff, supplies, facilities, equipment, and materials and the annual dollar amounts necessary to pay for or purchase them.
>
> *Activities*—what the program does. These may include marketing the program, recruiting clients, conducting individual or group meetings with clients, holding workshops for staff; and sponsoring conferences.
>
> *Outputs*—units of service or products. These include the number of clients and the number of hours devoted to providing services. The outputs, produced by activities, are expected to lead to outcomes.
>
> *Outcomes*—events, occurrences, or conditions representing an impact that the program has on the community outside of the organization. Outcomes can be immediate, intermediate, long term or ultimate.

To see how you can apply a logic model to a specific program, consider a program to enable low-income families to obtain affordable, adequate housing. Its long-term goals are to increase the number of low-income families in adequate, affordable housing and to reduce the number of families who are homeless or living in substandard housing. Figure 4.2 shows a logic model for the program.

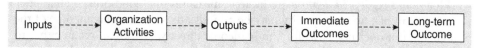

FIGURE 4.1
The General Structure of a Logic Model

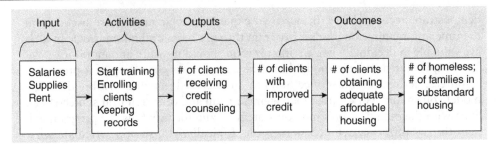

Input	Activities	Outputs	Outcomes		
Salaries Supplies Rent	Staff training Enrolling clients Keeping records	# of clients receiving credit counseling	# of clients with improved credit	# of clients obtaining adequate affordable housing	# of homeless; # of families in substandard housing

FIGURE 4.2
A Logic Model Applied to a Housing Program

Moving from the left to the right, think of each component as providing the condition for the next component. For example, what resources are necessary to operate the program? If you have these resources you can carry out the activities needed to meet the expected demand for services; if you provide the necessary services to families, participants will benefit by finding adequate housing.

Consider the program in Figure 4.2. To monitor activities and to assess how well the program is doing you would collect information on some components. A performance-monitoring system does not try to measure everything. You need to balance the benefits of obtaining information against the time and costs of doing so. Trying to obtain data on too many components may take too much time, drive up costs, and limit the quality of the data. Delivering a service typically requires several activities. For example, the housing program must recruit, notify, enroll, and counsel clients. You, along with a management team, must decide which activities are important to track to improve program operations and to help further the organization's mission.

In creating a logic model you may want to distinguish among short-term, intermediate, and long-term outcomes. Short-term and intermediate outcomes are those that should lead to a desired end but are not ends in themselves. Response times for medical teams are short-term outcomes expected to lead to the long-term outcome of reduced preventable fatalities. Other examples of long-term outcomes are an increase in the number of low-income families living in safe, affordable housing, a decline in juvenile crime, and a decrease in the incidence of disease.

Service quality is considered an intermediate outcome. Quality measures recipient perceptions about how well a service was delivered. Consider a program intended to prepare students to advance into leadership positions. Evidence of their successes may not be available soon enough to be measured and linked to the program. So students' perceptions about the quality of the program and how well they believe it will help them are treated as intermediate outcomes. Measuring these perceptions will help the program managers gauge if the program is working.

Three terms often associated with performance measurement are indicators, workload, and equity. *Indicators* are indirect measures of a condition or characteristic. For example, housing quality may be an indicator of the family income of the residents, but it does not measure family income directly. *Workload* refers either to activities or to outputs, such as the number of applications processed or the number of clients counseled, respectively. Workload data are valuable for planning, budgeting, and performance measurement. You should also be concerned about the clients receiving a program's services and eligible clients who are not. Uncovering the reasons for not receiving services provides valuable information. *Equity* refers to the fairness in providing program services. A widely held contemporary value is that social services should be available to all eligible persons without regard to race, ethnic group, religion, gender, or sexual orientation. For example, to demonstrate equity a food pantry may track the number of recipients by racial or ethnic group. You may want to go beyond showing that program services are widely available and used. Categorizing a program's outputs and outcomes by salient population characteristics may identify equity problems. The food pantry, for example, should be concerned if 50 percent of its recipients are Hispanic but only 10 percent of its recipients who adopt a nutritious diet are Hispanic.

Performance measures by themselves cannot confirm that a program was successful. Outcomes may improve because of other changes in the environment. For example, an increase in employment may be due to a better local economy, not the effect of job-training programs. Performance measures seldom provide answers to what should be done to improve the outcomes, for example, if nutritious diets are not adopted, if crime increases, or if disease spreads. Performance measures can, however, alert managers to problems and generate important discussions about what is causing disappointing outcomes.

EFFICIENCY AND PRODUCTIVITY

As a manager you will want to record on a regular basis the amount of resources used. You will probably also want to measure how well the program is using its allocated resources. For a food pantry, for example, you could record the total amount of food distributed each day. You could also determine the average cost of the food provided to each individual or family. As manager of a housing program you could measure the total cost of providing counseling to clients and determine the average cost per client. Since much of the cost of providing counseling is the time spent by counselors you might decide to determine the number of hours counselors spend per client. To assess your costs or time in providing services you would want to see how you are doing compared to earlier times or other programs. Typically managers compare their program to similar programs or to their program's previous performance.

Common performance measures for how well organizations or programs use resources are efficiency and productivity. *Efficiency* relates the amount of input to the amount of output or outcome. For example, the average cost (an input) of providing food to a family (an output) would be an efficiency measure. The average cost of improving a family's nutritional status (an outcome) would also be an efficiency measure. *Productivity*, on the other hand, is the ratio of the amount of output (or outcome) to the amount of input. The average number of clients counseled by each counselor per month would be a productivity measure. Efficiency and productivity have traditionally linked costs and outputs. A performance-monitoring system that provides data on outcomes gives a more useful picture of efficiency and productivity. Output-to-input ratios can be affected by increasing output at the expense of results (outcomes) and the quality of service. Consider job placement services. A service may report high efficiency because it counsels a large number of clients (output) but at the same time it may place people in low-paying, dead-end jobs (an outcome).

Efficiency and productivity are often used as indicators of performance. In general we would accept greater efficiency and productivity as indicators of better performance than lower efficiency or productivity. However, we may not be able to judge performance unless we can compare these to similar measures in other organizations or to our own in previous months and years.

Here are sample data that a housing program can use to measure and monitor performance. (Note how they link to the program logic model). We show here how the program manager can use this data to measure efficiency and productivity.

Inputs
   a. Cost of counseling clients in a 3-month period: $45,000
   b. Number of counselor hours devoted to housing program in a 3-month period: 768

Outputs
   c. Number of clients counseled in the housing program in a 3-month period: 203
   d. Number of clients who completed a credit counseling program: 47

Outcomes
   e. Number of persons who obtained affordable, standard housing: 68
   f. Number of persons whose credit scores increased after receiving credit counseling: 38

Output-Related Efficiency Measures

Cost of counseling clients/number of clients counseled
   = $45,000/203 = $221.67 per client counseled

Time spent counseling clients/number of clients counseled
   = 768/203 = 3.8 hours per client counseled

Outcome-Related Efficiency Measure

Cost of counseling clients/number who obtained housing
   $45,000/68 = $661.76 per person obtaining housing

Outcome-Related Productivity Measure

Number who obtained housing/cost of providing program
   = 68/$45,000 = 0.00151 or, 1.51 clients obtained housing
   for each $1,000 of program cost

In addition to monitoring the amount of resources and the number of clients counseled, managers develop indicators of how well their programs convert resources into outputs or outcomes. Managers might track in 3-month intervals the unit cost of counseling clients and the unit cost of counseling clients who obtained housing. Together these indicators allow managers to monitor whether they are using more or less resources than planned and if the average cost of serving each client is changing.

DATA SOURCES

Because performance monitoring requires regular data collection you must pay attention to data quality. You want to make sure that the measures are reliable or you risk incorrectly comparing two time periods or two programs. Also, you can't ignore the cost of collecting, compiling, storing, and analyzing data. You may have to rely on other staff members to collect the data. They should be convinced that the data are worthwhile, otherwise they may report incomplete or erroneous data. In the following section, we identify five common sources of data to monitor performance. For the most part each data source has the added advantage of having a value beyond just monitoring performance.

Administrative Records

Organizations and programs routinely keep data on customers and clients. The records contain information on inputs (e.g., costs), outputs (e.g., services provided), and key client demographics (e.g., age, gender, ethnicity.). The records may contain information from staff members who submit regular performance reports, for example, the number of clients served, the types of services offered, and whether particular clients will continue to receive services.

Agency records have the advantages of availability and low cost. However, the information may not be easily converted into performance measures; the records may contain little or no information on quality; they may not indicate when the service was delivered; and a program manager's access may be limited because of confidentiality.

Customer Feedback

User surveys, focus groups, and interviews are all used to learn about customers' experiences with a program and their level of satisfaction. A user survey contacts samples of customers or clients. Respondents can provide data to measure outputs (services received), outcomes, and satisfaction level. Households in a community may be surveyed to identify potential or actual demand, which allows stakeholders to assess whether a program's outputs and outcomes are adequate.

Trained Observers and Rating Systems

Trained observers may use rating systems to assess outcomes that are more qualitative. For example, a community revitalization program may create rating scales and have raters assess the cleanliness of streets, the appearance of yards, and the physical condition of residences.

Census Data

The U.S. Census Bureau records provide useful information on community characteristics that can be used for performance information. The decennial census (completed every ten years) includes demographic information on the following: ethnicity, age, and gender; the number of homeowners and renters; and household size. More current and extensive information is included on the American Community Survey conducted every year.[2] Each state has a Data Center, created as a partnership between the Census Bureau and the state government, which provides users with access and education on census data and other statistical resources.[3]

Government Records

State and local governments are also potential sources of data. Counties record information such as births, deaths, and chronic diseases for their jurisdictions and pass this information on to the state. State government Web sites typically describe various sources of state and local statistical data and have links to these sources. The data available includes existing and projected population numbers, other demographic information, housing, maps, and much more.[4]

TIME SERIES

Regular data collection and tracking are key characteristics of performance measurement. The more frequently data are gathered the higher the cost. This is especially true of surveys or rating systems which can be quite costly each time the data are gathered. The drive for accountability and performance information for budget planning seems to favor annual reporting. More frequent reporting may occur if a measure lends itself to frequent review and adjustment, that is, as part of continual program improvement.

Quantitative information collected regularly over a long time constitutes a time series. The data can depict both short-term changes and long-term trends in a variable. Most people are familiar with time series that regularly report on some aspect of the nation's economy or a social condition. The unemployment rate, crime rate, number of people below the poverty level, and number of new homes sold are examples. Time series are suited for situations when you want to

- describe changes over time;
- monitor trends;
- forecast future trends;
- estimate the impact of a program or an intervention;
- establish a baseline for future comparisons.

Housing program managers will probably want to track the number of clients that are enrolled or seen each month, each quarter—every three months—and every year. Since goals of the organization include reducing the number of homeless families and the number living in substandard housing, program managers will want regular information on these indicators for the community to see if they have

---

[2]http:www.census.gov/; http:www.census.gov/acs/www/

[3]The State Data Center Program is described at http://www.census.gov/sdc/

[4]For example, a North Carolina State Government site provides a Web resource for North Carolina data with over 1,300 data items from state and federal agencies. See http://linc.state.nc.us/

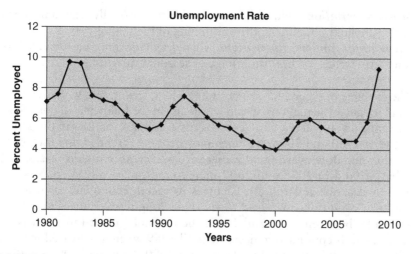

FIGURE 4.3
National Unemployment Rate: 1980–2009

changed. With this information managers document changes in the need for services and plan for future services. Time series data presented in graphs and tables are generally easy to interpret and are useful in routine decision making. For example, managers of the housing program may want information on the number of clients enrolled and served and the length of service time to help them allocate resources. A time series might include an explicit independent variable. For example, housing office staff might analyze a time series of the number of people in need of adequate housing at various times before and after an event such as a major area employer going out of business. The employer's closing would represent the independent variable and the number of people needing housing the dependent variable.

The data are typically presented in a graph with time along the horizontal axis (*X* axis) and a variable's values along the vertical axis (*Y* axis). You should focus on the variable and its changes or variations over time. Figure 4.3 shows the annual unemployment rate over nearly 30 years. This example reports data for a sample of U.S. households. You could develop a similar graph for a single state, county, or city sample and for different time periods.

Time series are useful to show the changes in a variable. As you look over the graph in Figure 4.3, look for patterns—dips and rises—in changes in the unemployment rate. Four types of changes or variations occur within a time series:

1. *Long-term trends.* General movement of a variable, either up or down, usually for 5 or more years.
2. *Cyclical variations.* Regular up-and-down changes in a variable that occur within a long-term trend; one- to five-year cycles are common although longer cycles are often observed.[5]

---

[5]M. Kendall and J. K. Ord. *Time Series*, Third Edition (New York: Oxford University Press, 1990). Kendall and Ord identify four-year cycles. Business cycles are often between three and eight years.

3. *Seasonal variations.* Fluctuations traceable to seasonally related phenomena such as holidays, weather, or the beginning of the school year.
4. *Irregular (or random) fluctuations.* Changes that cannot be attributed to long-term trends, cyclical variations, or seasonal variations.

The example in Figure 4.3 demonstrates a long-term trend, with cyclical variations and irregular fluctuations, which we discuss in the following section. Since the data are annual you will not be able to identify seasonal variations in this series.

Between 1980 and 2006, the unemployment rate trended downward. However, within that downward trend we see several increases and decreases. In 2006 it leveled off and in 2008 and 2009 increased dramatically. Variables seldom increase or decrease indefinitely. Changes in social, economic, environmental, and political conditions may lead to short- or long-term changes in a variable. For example, between 1980 and 1985, the unemployment rate increased and then decreased. It continued to decrease until 1989 when it increased for 3 years, then decreased, and continued to do so until 2000. Between 2000 and 2005 we again see an increase and decrease and then the dramatic increase in 2008 and 2009. When we started writing this text we couldn't tell if the 2008 spike was an irregular fluctuation or if it signaled a longer period of relatively high unemployment. Now in 2010, the 2008 change at best marks a cyclic shift in direction or at worst the beginning of a long-term trend. By the time you read this you may have additional information and will be in a better position to judge.

Cyclical variations are regularly occurring fluctuations within a long-term trend that last for more than a year. A complete cycle is from "peak to peak" or "valley to valley" in a time series. Sometimes cycles are very regular. For example, the percentage of Americans voting peaks every 4 years, during presidential elections. The variations illustrated by the unemployment rate are more commonly seen than are regular cycles. Up- and- down-movements take place throughout a time series but the cycles may not be of the same length and may last longer than 5 years. The unemployment data for the period 1980–2005 have two cycles of approximately 10 years each. You see long stretches when the unemployment rate decreased. Between 1980 and 2005 periods of increasing unemployment rarely lasted more than 3 years. Perhaps political interventions brought about a change. For those who use advanced statistical tools to analyze time series and use them to forecast, regular cycles are necessary. However, program managers can obtain useful information from understanding that these changes take place. Awareness of these variations can help the manager avoid incorrectly attributing changes to an action of the organization or in concluding that a long-term trend has changed.

Seasonal variations describe changes that occur within the course of a year. Data must represent time intervals that recur within a year, such as days, weeks, months, or quarters (3-month periods). The observed fluctuations occur within a single year and recur year after year. If the data in Figure 4.3 were monthly, you would have seen several fluctuations in any given year. Such changes are considered seasonal variations only if a similar pattern is seen year after year. An administrator may use seasonal information to decide how to staff facilities—how many staff to hire for city parks and for what length of time, for example, for 10, 12, or 14 weeks. A program manager administering an emergency shelter that secures temporary housing for homeless men would know that more individuals will need housing during certain times of the year. Ignorance of seasonal demands may result in erroneous decisions. Imagine the consequences if a merchant assumed that December sales marked a business upswing that would carry through the next several months.

Irregular fluctuations are variations not associated with long-term trends, cyclical variations, or seasonal variations. The irregular variations may be nonrandom or random. A nonrandom fluctuation results from conditions that can be identified and that explain the variation. The conditions may be inferred from records of concurrent events. For example, a news story about an emergency shelter or the occurrence of a flood or other natural disaster may explain an increase in homeless persons seeking shelter. Random movements are those that seemingly occur with no explanation that can be found. They are unpredictable and tend to be relatively minor

To infer that a program made a difference you might compare the difference in the time series before and after the program was implemented. You might look at a time series to help determine the impact of a major event, such as a flood, an earthquake, or an employer closing. You might consider outcomes such as changes in unemployment, the number of homeless persons, and evidence of psychological depression.

Social indicators quantify social conditions and are often presented as a time series. The unemployment rate shown in Figure 4.3 tracks a nationwide social condition. The infant mortality rate shown in Figure 4.4 tracks a statewide social condition. (You should be able to identify the long-term trend and some cyclical variations in this time series.)

Leaders of nonprofit organizations and government policy makers may not be able to affect social conditions in a short time. However, the time series can alert them to serious problems, and awareness of possible changes in the time series will help them gauge if changes are likely to be long term.

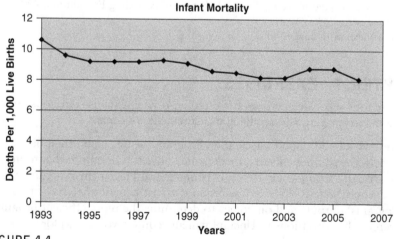

FIGURE 4.4

Infant Mortality Rate in a Large State: 1993 to 2006 (Number of Deaths per 1,000 Live Births)

If government leaders in the state recording the data in Figure 4.4 were concerned about the long-term level of infant mortality, they would be relieved that it appears to be declining. However, during the time span shown the mortality rate increased before continuing its decline. When an increase is observed officials can investigate reasons for the increase and take appropriate actions.

CONCLUDING OBSERVATIONS

This chapter began with a discussion of how organizations use logic models to develop programs and to provide a framework for measuring program performance. The components of the logic models identify steps in delivering service and achieving desired ends. These components help managers identify how to monitor the use of resources, organizational activities, and their results. Logic models link program planning and performance monitoring. The chapter identified the types of measures used to monitor performance and where to find data.

The chapter concluded by describing time series, a common strategy for using performance data. Time series can track the use of resources, activities, outputs, and outcomes. Understanding the variations in time series can enable you to respond to changes in a program and its environment. As we will explain in Chapter 10 a time series cannot demonstrate that a program caused observed changes in social conditions. Chapter 5 turns to a discussion of the statistics used to analyze the data collected as organizations and programs monitor performance.

RECOMMENDED RESOURCES

Hatry, Harry, Phil Schaenman, Donald Fisk, John Hall, Louise Snyder. *How Effective Are Your Community Services? Procedures for Performance Measurement*, Third Edition (Washington, D.C.: The Urban Institute Press, 2007).

Urban Institute, Outcome Indicators Project, http://www.urban.org/center/cnp/projects/outcomeindicators.cfm

The Urban Institute and The Center for What Works have an outcome indicators project with suggested outcome indicators for 14 social programs. The information is available at http://www.urban.org/center/cnp/projects/outcomeindicators.cfm

W. K. Kellog Foundation, *Logic Model Development Guide: Using Logic Models to Bring Together Planning, Evaluation, and Action* (Battle Creek, Michigan: W. K. Kellog Foundation, 2004)

CHAPTER 4   EXERCISES

There are four separate exercises for Chapter 4. Each exercise develops your competence in interpreting and applying measurement concepts.

- Exercise 4.1 Childhood Vaccination Program focuses your attention on logic models, the components of a model, and choices in deciding what to measure.
- Exercise 4.2 Comparing Two Job Training and Placement Programs focuses on cost efficiency. This exercise requires a cost benefit analysis of two programs' outputs and outcomes to determine the more efficient program.
- Exercise 4.3 Variations in Unemployment requires you to explore and interpret time series data.
- Exercise 4.4 On Your Own provides an opportunity to work with these concepts on your own or in your own agency.

EXERCISE 4.1 Childhood Vaccination Program

Scenario

The Mid Valley Family Health Center is launching an aggressive antiflu campaign in the communities served by its clinics. The campaign will consist of two major activities: a publicity campaign about behaviors that can prevent the spread of flu and providing flu shots at minimal or no cost. The clinics, which serve low-income households, are located in racially and ethnically diverse communities.

Section A: Getting Started

1. Write a long-term outcome for this initiative.
2. Create a logic model for this initiative. (You may present the two activities on one model or separate models.) Be sure to include information about the inputs, activities, outputs, and outcomes for the activities.
3. Suggest three benefits of drawing a logic model.
4. Consider the plan to provide flu shots and suggest how you would measure
   a. input
   b. activity
   c. output
   d. impact
   e. workload
   f. program quality
   g. equity.
5. Consider the plan to provide flu shots. What would you need to measure and how often would you collect the data for the following purposes?
   a. To estimate how many staff you need at each site
   b. To determine if program participation is equitable
   c. To estimate your budget for next fiscal year
   d. To include in your annual report to the board of directors

Section B: Small Group and Class Exercise

In groups of three to five prepare a presentation of the logic model to the program initiative's management team.

1. Compare your logic models (Section A) and decide on a model for your presentation.
2. Prepare your presentation.
   a. Comment on the value of the logic model to the Health Center and its staff.
   b. Identify the long-term outcome and how it might be measured.
   c. Suggest what components you would suggest that the Health Center track
      i. at least quarterly
      ii. at least annually

d. For what quality indicator(s) would you suggest gathering data? Justify its (their) importance.

e. For what equity indicator(s) would you suggest gathering data? Justify its (their) importance

Class Exercise

- One group should make its presentation to the class.
- One or two groups should play the role of the management team and ask questions for clarification, to identify challenges or problems with the model or with implementing the plans for monitoring performance. The "management team" should point out strengths of the proposal.
- All other students will act as observers and assess each group's performance, focusing on those aspects of the presentation, questions, and answers that were particularly compelling.

EXERCISE 4.2   Comparing Two Job Training and Placement Programs

Performance information can be used to measure efficiency. The ratio of the amount of input to the amount of output is efficiency, and the ratio of the amount of output to the amount of input is productivity. For this exercise calculate the efficiency in terms of dollar costs.

1. Consider the job training and placement programs administered by two different organizations. The programs' respective output, outcome, and input measures are given below. Calculate the output and outcome efficiency of the program for a 3-month period.

    Output measures
    a. Number of clients who received job training and placement services in a 3-month period (833)

    Outcome measures
    b. Number of persons who obtained a job (616)

    Inputs
    c. Number of dollars expended to provide services to clients in the job training and placement program in the 3-month period ($150,000)

2. Next let's compare this job training program (Program A) to another program (Program B) with the same goals. Calculate the efficiency of Program B, whose output, outcome, and input measures are as follows.

    Output measures
    a. Number of clients who received job training and placement services in a 3-month period (150)

    Outcome measures
    b. Number of persons who obtained a job (68)

    Inputs
    c. Number of dollars expended to provide services to clients in the job training and placement program in the 3-month period ($50,000)

3. Identify the more cost-efficient program.

4. Explain how efficiency might be different if the target population for Program A was welfare mothers and for Program B was graduating high school seniors.

5. How may the quality of the job affect program efficiency? Consider that jobs can range from low-skilled temporary jobs to skilled permanent jobs.

EXERCISE 4.3   Variations in Unemployment

Generally job training programs are popular during times of high unemployment. Job training programs use the unemployment rate to justify services, particularly increases in services. The following table contains the U.S. unemployment rate data from January 2006 to December 2009. How would you arrange these data in a time series? Graph these data. Consider putting months on the X axis and creating a separate line for each year.

Year	Jan	Feb	Mar	Apr	May	Jun	Jul	Aug	Sep	Oct	Nov	Dec
2006	4.7	4.8	4.7	4.7	4.7	4.6	4.7	4.7	4.5	4.4	4.5	4.4
2007	4.6	4.5	4.4	4.5	4.5	4.6	4.7	4.7	4.7	4.8	4.7	4.9
2008	4.9	4.8	5.1	5.1	5.5	5.6	5.8	6.2	6.2	6.6	6.8	7.2
2009	7.7	8.2	8.6	8.9	9.4	9.5	9.4	9.7	9.8	10.1	10	10

Source: U.S. Department of Labor, Bureau of Labor Statistics, Labor Force Statistics from the Current Population Survey.

Write a brief statement describing these data as they are depicted in the graph you have created. Be sure to discuss any long-term trends, cyclical variations, seasonal variations, or irregular (or random) fluctuations. For any random fluctuations, can you suggest economic, political, or social factors that may have impacted these changes? How could the job training program use these data?

EXERCISE 4.4   On Your Own

Identify a program at your job or internship.

1. Based on a review of program materials and staff interviews, draw a logic model. Focus on
   a. long-term objectives.
   b. the activities conducted to achieve objectives.
   c. the input needed to achieve the objectives.
2. Review your model with a staff member and get his or her feedback as to its accuracy, clarity, and usefulness.
3. Make a recommendation for measuring program quality.
4. Make a recommendation for measuring program equity.
5. What measures does the agency currently track about the program? Why are these collected?
6. What data should be collected to demonstrate effectiveness? How often should these data be collected?
7. Use data that your agency collects on a regular basis. Create a chart, graph these data, and describe the fluctuations

Analyzing Performance Measures

Think of data as the raw materials that you convert into information. Data by themselves, however, are not likely to be very useful. Column after column of numbers mean very little. To make the data meaningful you need to organize and present them, that is, the data need to become information. As you skim a newspaper, listen to a presentation, or read a report you may mistakenly assume that creating the graphs and statistics took little work. This is not true. Someone thought through how to present the data so that you and others could quickly understand and interpret them.

One of your tasks as a program manager is to decide how to organize and present data. There is no single best way to graph or analyze a set of data. You may create several graphs and try different statistics as you search for patterns that make sense. In this chapter we focus on the basic tasks for organizing performance data: entering data into a spreadsheet, creating tables and graphs, and describing variations in individual variables. The same skills apply to surveys, program evaluations, and community assessment, which we cover in later chapters. Also note that in Chapter 8 we will discuss analyzing relationships between and among variables. First, however, we will cover the terminology of measurement scales. Familiarity with these terms will facilitate our discussion of various statistics in this chapter and later.

MEASUREMENT SCALES

Measurement scales or levels of measurement describe the relationship among the values of a variable. You will find the terminology associated with measurement scales useful as you decide what statistics to use. The basic scales are nominal, ordinal, interval, and ratio scales.

Nominal scales identify and label the values of a variable. You cannot place the values of a nominal variable along a continuum; nor can you rank individual cases according to their values. Even though numbers are sometimes assigned, these numbers have no particular importance beyond allowing you to classify and count how many cases belong in each category. For example, imagine an organization, the Happy Housing Center, records why people seek its services. The variable Reason for Seeking Services has four values: "Laid off or lost job," "Rental housing needs repairs," "Rent increased," "Eviction." A nominal scale reports how many requests for services are in each category:

1 = Laid off or lost job

2 = Rental housing needs repairs

3 = Rent increase

4 = Eviction

The numbers are simply a device to identify categories; letters of the alphabet or other symbols could replace the numbers and the meaning of the scale would be unchanged. Remember, too, that values of nominal scales are not ranked. Thus, the numbering system in our example does not imply that an eviction has a greater or lesser value than being laid off.

Ordinal scales identify and categorize values of a variable and put the values in rank order. Ordinal scales rank the values without regard to the distance between values. Ordinal scales report that one case has more or less of the characteristic than another case does. If you can rank values but do not know how far apart they are, you have an ordinal scale. You assign numbers to the values in the same order as the ranking implied by the scale. For example, the value represented by 3 is greater than the value represented by 2, and the value represented by 2 is greater than the value represented by 1. The numbers indicate only that one value is more or less than another; they do not imply that a value represents an amount. Let's look at how you could assign numbers to respondents' answers to the statement "The Happy Housing Center staff provided me with accurate information."

5 = Strongly agree

4 = Agree

3 = Neither agree nor disagree

2 = Disagree

1 = Strongly disagree

You can use other numbering schemes as long as the numbers preserve the rank order of the categories. For example, you could reverse the order and number Strongly agree as 1 and Strongly disagree as 5. Alternatively, you could skip numbers and number the categories 10, 8, 6, 4, and 2. Because you cannot determine the distance between values, you cannot argue that a client who answers "Strongly agree" to all items is five times more satisfied than a client who answers "Strongly disagree" to all items.

Rankings commonly produce an ordinal scale. For example, a supervisor may rank 10 employees and give the best employee a 10 and the worst a 1. The persons rated 10 and 9 may be exceptionally good, and the supervisor may have a hard time deciding which one is better. The employee rated with 8 may be good, but not nearly as good as the top two. Hence, the difference between employee 10 and employee 9 may be very small and much less than the difference between employee 9 and employee 8.

Interval and *ratio* scales assign numbers corresponding to the magnitude of the variable being measured. Interval scales do not have an absolute zero; ratio scales do. The most common example of an interval scale is the temperature scale. We know that 40°F is 20° warmer than 20°F, but we cannot say that 40° is twice as warm as 20°. In the Fahrenheit temperature scale, zero is an arbitrary point; heat exists at 0°F.

The numbers you assign to a ratio scale could be the actual number of persons working in an agency, the number of homeless persons in a given year, or the amount of per capita income in a city. You can add or subtract the values in ratio scale. If the Happy Housing Center had 100 service requests in January and 50 in February, you can say that the number of requests fell by 50 in February. You may also note that the center had half as many requests in February. And at zero, there are no service requests. Table 5.1 summarizes information on these four levels of measurement.

In practice, the boundaries between ordinal and interval scales and interval and ratio scales may be blurred. If an ordinal scale has a large number of values, analysts may assume that it approximates an interval scale. Similarly, the summed values from a set of questions, such as the six questions that measured orientation quality in Chapter 2, may be treated as a ratio scale.

You may mistakenly assume that categories consisting of numerical data form a ratio scale. They do not. Rather, such a scale would be ordinal. For example, the following categories constitute an ordinal scale: under 20 years old, 20 to 29 years old, 30 to 39 years old, and so forth. The exact distance, that is, the age difference between any two people, cannot be determined. While we know a person who checks off 20 to 29 years old is younger than a person who checks off 30 to 39 years old; the age difference between them could be a few days or nearly 10 years.

TABLE 5.1

Levels of Measurement

Characteristic	Nominal	Ordinal	Interval	Quantitative
Cases assigned to categories	Yes	Yes	Yes	Yes
Cases ordered according to value of variable		Yes	Yes	Yes
Values refer to amounts for cases			Yes	Yes
All units of measurement scale are equal			Yes	Yes
Zero on scale indicates no value for variable being measured				Yes

Examples:

Nominal
  Region of the country: East, South, West, North

Ordinal
  Rating of nonprofit organization on aid to poor: Very high, High, Average, Low, Very low

Interval
  Fahrenheit temperature on volunteer workday: 55°

Ratio
  Number of days per month shelter has no open beds: 5

ENTERING DATA ON A SPREADSHEET

The first step in organizing data is to enter them into a spreadsheet. A spreadsheet may be a piece of paper with rows and columns or a software tool, such as Microsoft Excel. When you open an electronic spreadsheet a workbook with rows and columns appears on the computer screen. Each column can represent a variable and each row can represent a case. You can enter numbers or text into each cell. Entering data is relatively easy. In addition to storing data most spreadsheet programs can calculate basic statistics and create graphs. The graphs can be moved into text documents as part of a report. The data can be readily transferred into a sophisticated statistical software package for more intense analysis.

Before you enter the data you should have a plan for how you will use them. You will want to decide whether to enter words or letters as opposed to numbers. Statistical analysis is easier with numbers; but words and letters are easier to understand. To illustrate other common decisions consider the data on Robert and Anne, two clients who visited a Happy Housing Center. The counselor who worked with them recorded the following.

Name	Gender	Age	Reason for Requesting Services
F. Robert	Male	54	Needs Help Obtaining Apartment
P. Anne	Female	39	Received Eviction Notice

The names and the values of three variables, with the exception of age, could be entered as text, but statistical analysis may be easier if numbers represent the values for gender and reason for requesting services. A "1" may be entered for Males and a "2" for Females; "1" may be assigned to Needs Help Obtaining Housing, "2" for Received Eviction Notice, and so on. If numbers were used, the information for Robert and Anne would look like this:

Name	Gender	Age	Reason for Requesting Services
F. Robert	1	54	1
P. Anne	2	39	2

You can decide what numbers to assign to a value. Instead of "1" for males and "2" for females, other numbers could be used. If the values constitute an ordinal scale, the numbers assigned should correspond to a continuum. Responses ranging from Very satisfied to Very dissatisfied could just as easily been "5" for Very dissatisfied and "1" for Very satisfied.

Data for a given case may be incomplete. Data in existing records may be missing, a respondent may have refused to answer a question, or some items on a form may not be legible. A missing value may be represented by "missing," or "not applicable," or a 9, 99, or another number that is noticeably larger than the other values. A large number may enable you to easily identify and track missing data. It also reduces the risk of including missing data in statistical calculations, because a

9 or 99 will often yield a statistical result that doesn't make sense. For example, assume that the Housing Center also recorded the number of times a client had been evicted. If the arithmetic average for this turned out as a high number, say 65, then you would know something was wrong. However, if a missing answer for this variable was entered as 0, then you might not notice that this value was included in calculating the mean.

COUNTING THE VALUES: FREQUENCY DISTRIBUTIONS

The first step in any analysis is to obtain a frequency distribution for each variable. A frequency distribution lists the values or categories for the variable and the number of cases with each value. For example, a frequency distribution of gender in the database containing Robert's and Anne's information would report the number of males and females. A frequency distribution for age would report the number of respondents who were 22, 23, 24 years old, and so on. More likely you would combine the age categories, for example, less than 20 years old, 20–29 years old, 30–39 years old, and so on. The categories must be set so that you count each case in one and only one category.

Some variables lend themselves to grouping. If the differences between some groups' values are of little interest, you may combine responses. For example, you may combine "Strongly Agree" with Agree and "Strongly Disagree" with Disagree. A variable, such as income, may have so many values that you cannot easily discern how it varies without grouping the values in some way. Presenting each value of income separately may result in a large number of values, most of which will have few, if any, cases. Therefore, you need to decide how many categories to create and how wide each category should be. The categories should not be so wide that important differences are overlooked. Nor should they be so narrow that many intervals are required. Using equal intervals to group ages, by decades as shown in Table 5.2, is a common way to create categories, but for some variables equal intervals may mask important differences among cases. A frequency distribution for income will often have many cases in the lower income categories and few in the higher income categories. To give a more accurate picture you may want to create narrower categories at the lower income levels and wider intervals at the higher income levels.

TABLE 5.2

Frequency Distribution of Ages of Happy Housing Center Clients

Age (in Years)	Number of Cases
Less than 20	2
20–29	6
30–39	12
40–49	22
50–59	20
60–69	17
70–79	3
Total	82

Relative Frequency Distribution

Relative frequency distributions report the percentage of cases for each value. (A *percent* is calculated by dividing the frequency of one value of the variable by the total number of cases and multiplying the result by 100.) The percentages allow you to compare the frequency of different values in the same distribution or to compare values in two or more frequency distributions that have different numbers of cases. Almost everyone is familiar with percentages and can quickly interpret them. In our experience politicians and the public are comfortable with findings that are either preceded by a dollar sign or followed by a percent. So don't hesitate to report percentages and make them the focus of a presentation.

You can report frequency distributions and relative frequency distributions in the same table. Uncluttered tables are easier to read; therefore, you may want to report the relative frequency for each value and include enough information so that a reader can determine the number of cases represented by each percent. You may report the total number of cases as a column total as in Table 5.2, in the table title or a part of the row or column label as in Table 5.3.

Cumulative Relative Frequency Distribution

You may want to show more than just the percentage of cases in a specific category, or you may want to indicate the percentage of the cases below a given value. A cumulative relative frequency distribution provides this information. To obtain the cumulative percentage, the percentages up to the given value are added together. Table 5.3 includes the relative frequency distribution and cumulative percentage distribution for the data in Table 5.2. The percentage of cases column gives the total number of cases, so you can multiply the total number of cases by the percentage to learn how many cases are found in each category. Remember to convert the percentage to a decimal before multiplying. The last column reports the cumulative percentage for the distribution of ages. Note, for example, that 24.4 percent of the clients seeking housing assistance were under the age of 40. Note also that the Number of Cases column of Table 5.2 is omitted from Table 5.3.

TABLE 5.3

Distribution of Client Age in Happy Housing Center

Age (in Years)	Percentage of Cases ($N = 82$)	Cumulative Percentage
Less than 20	2.4	2.4
20–29	7.3	9.7
30–39	14.6	24.4
40–49	26.8	51.2
50–59	24.4	75.6
60–69	20.7	96.3
70–79	3.7	100.0

Presenting Data Visually

Once you know the frequencies, you need to decide how to present them. Visual presentations of data often illustrate points more clearly than do verbal descriptions. Some people who are not comfortable with tables and statistics seem to have no trouble interpreting a well-done graph. Visual displays[1] should

serve a clear purpose; to describe, explore, or elaborate;

show the data and make them coherent;

encourage the eye to compare different pieces of data;

entice the reader or listener to think about the information;

avoid distorting what the data have to say;

enhance the statistical and verbal descriptions of the data.

A graph should be able to stand on its own. You should give each graph a descriptive title and label its variables and their values. Use footnotes to clarify any terms that need an explanation to be interpreted correctly and to identify the source of the data. You will find it useful to indicate the date when a graph was produced because, as you analyze the data, you may create multiple versions as you correct errors and make changes in how you combine values. If graphs are not dated you can lose time trying to determine which version is the most recent.

Pie Charts

A *pie chart* consists of a complete circle, or pie, with wedges. The circle represents 100 percent of the values of the variable displayed. The size of each wedge or "slice of the pie" corresponds to each value's percentage of the total. Figure 5.1 depicts a pie chart showing the primary reason why people requested services from the Happy Housing Center.

A common convention is to place the largest slice of the pie at the 12 o'clock position. The other slices should follow in a clockwise direction according to size, with the smallest slice last. This rule, however, does not apply if you create two pie charts to make a comparison, for example, if you want to compare the reasons for requesting services from the Happy Housing Center last year and 10 years ago.

Pie charts work well for oral presentations and short written reports, because an audience can quickly understand the depicted information. They may effectively illustrate differences among organizations, locations, and dates. You should avoid using a pie chart that requires a large number of slices. An audience may find it difficult to differentiate between the slices, especially if it is comparing two charts. Furthermore, with many slices you may have trouble finding distinctive colors or patterns to distinguish one slice from another.

---

[1]Edward R. Tufte, *The Visual Display of Quantitative Information* (Cheshire, CT: Graphics Press, 1983), p. 13.

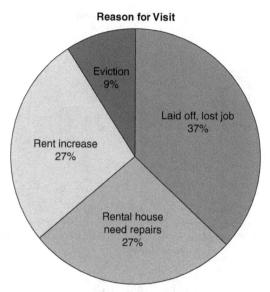

Reason for Visit

Eviction
9%

Laid off, lost job
37%

Rent increase
27%

Rental house
need repairs
27%

FIGURE 5.1
Reason for Visiting Homeless Office – Pie Chart

Bar Graphs and Histograms

Bar graphs are alternatives to pie charts. You place the value of the variable along one axis and the frequency or percentage of cases along the other. The length of the bar indicates the number or percentage of cases possessing each value of the variable. Figure 5.2 depicts a bar graph that includes the same data as the Figure 5.1 pie chart.

Whether the bars are vertical or horizontal depends on which arrangement communicates more effectively and clearly. All bars in a graph should have the same width. If you use different widths for the bars, you risk implying that some values are more important than others, which is misleading. You can also use bar charts to compare organizations, locations, or dates. For example, if you had historical data you could compare this year's data in Figure 5.2 to data from 10 years ago. To do this you would place a bar representing the percentage laid off 10 years ago next to the first column. The percentage whose rental house needs repairs would go next to the second column and so on.

A *histogram* represents ratio variables. Figure 5.3 shows an example of a histogram. Each column of the histogram represents a range of values; for example, in Figure 5.3 the second column represents the range 20–24 years old, and the third column the range 25–29 years. The columns adjoin one another because the range of values is continuous. The variable and its values are displayed along the horizontal axis, and the frequency or percentage of cases is displayed along the vertical axis. A histogram is similar to a bar graph, but unlike bar graphs its widths can vary. For example, the widths would vary if the age groupings for the columns were as follows: less than 20, 20–24, 25–29, 30–44, 45–59, and over 60.

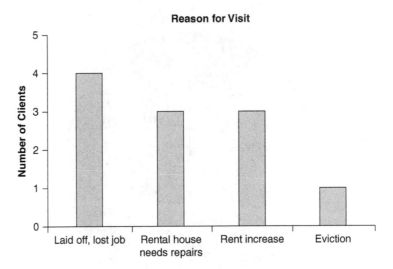

FIGURE 5.2
Reason for Visiting Homeless Office – Bar Chart

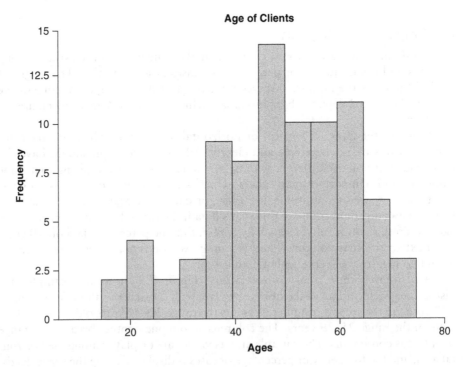

FIGURE 5.3
Age of Clients of Homeless Office

Time Series Graphs

To track performance over time you will want to use time series graphs. They are valuable to monitor performance, show changes , and demonstrate the impact of a policy. Users can easily and quickly discern changes from one time period to another, trends over time, and the frequency and extent of irregular fluctuations. (Recall that Chapter 4 discussed the changes over time you should look for.) To create a time series graph you put time, whether it is days, months, or years, on the horizontal axis, and the values of the variable on the vertical axis. For each time period place a dot at the intersection of the time period and the variable's value, and draw a line to connect the dots. Figure 5.4 shows an example of a time series graph.

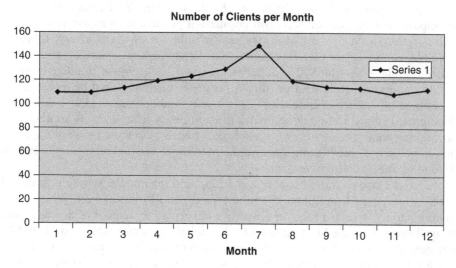

FIGURE 5.4
Number of Clients by Month

RATES AND PERCENTAGE CHANGE

We were tempted to title this section "putting your elementary school math to use." Calculation of rates and percentage changes requires only basic math skills. Both rates and percentage changes contain valuable information that policy makers and the public can understand and react to.

Rates

Rates report the number of cases experiencing an event as a proportion of the number of cases that could have experienced that event over a specific time period. Commonly reported rates include the unemployment rate and the crime rate. The unemployment rate reports the number of unemployed individuals as a percentage of the number who could have been employed (employed + unemployed). (The discerning reader will note that the number who "could have been employed" must be carefully defined. One common definition includes the "employable population actively looking for work." This would exclude those who are not seeking employment.) You may report rates as percents or use a base number other than 100. For example, cities, states, and nations report the annual rates of violent crimes as the number of occurrences for every 1,000 residents.

Assume that a county agency wants to compare the extent of homelessness in its community with that of other jurisdictions. Knowing the number of homeless may be valuable, but knowing how the problem in large cities compares with that in small cities or suburbs is also important. Rates allow such comparisons even though the cities may vary greatly in size.

An important decision is what to put in the denominator, that is, the number who could have been homeless. For many rates, the denominator is population size, but not always. As our definition of *rates* implies, the denominator for the unemployment rate excludes certain population groups, such as the very young. The selection of the denominator may appear to be somewhat arbitrary. Take, for example, contraceptive use. To compare data on contraceptive use by putting the entire population—which also includes men and children—in the denominator would not give an accurate a picture. A far better method would be to use either the number of women of childbearing age or the number of married women. Either denominator more accurately estimates the number at risk; *at risk* is another way of thinking about the number of possible occurrences. Deciding between the number of women of childbearing age and the number of married women may largely depend on the availability of data.

Consider two counties Moburg and Robus and the number of infant deaths in each. In one year Moburg had 104 infant deaths and Robus had 20. Which community had the greater problem? Directly comparing the number of deaths would be misleading because Moburg has 511,400 inhabitants and Robus has 106,000. Dividing the frequency of infant deaths in each county by the county population produces a more useful comparison. For Moburg we divided 104 by 511,400 which equals 0.0002033. For Robus, we divided 20 by 106,000 and obtained 0.00018886. The decimal values are so small that they might be ignored or interpreted incorrectly. Multiplying each decimal by 10,000 converts the data into figures that are more easily understood. We would report that Moburg has 2.033 infant deaths per 10,000 inhabitants and Robus has 1.886 infant deaths per 10,000 inhabitants.

The equation to compute a rate is

$$\text{Rate} = \frac{N_1}{N_2} \times \text{Base number}$$

where
N_1 = count for variable of interest
N_2 = population or another indicator of number of cases at risk
Base number = a multiple of 10

The following conventions apply in selecting a base number. Remember that these are conventions, not absolute rules.

- Be consistent and report rates in common use for a specific variable. Crime rates, for example, are usually reported as crimes per 1,000 of the population. Homicide rates, however, may be reported per 100,000 of the population.
- Select a base number that produces rates with a whole number with at least one digit and not more than four digits.
- Use the same base number when calculating rates for comparison. For instance, in the example just mentioned, you should not use a base number of 1,000 for Moburg and a base number of 10,000 for Robus.
- Note that a rate is meaningful only if it is specified for a particular time period, usually a year.

As noted earlier, you should also consider whether the entire population is the appropriate denominator for a rate

Percentage Change

The *percentage change* measures the amount of change over two points in time. For example, organizations in Moburg County have a campaign to reduce homelessness every year over the next decade. If they know the number of homeless people in any two years, they can report the change as a percent.

The formula for percentage change is

$$\text{Percentage change} = \frac{N_2 - N_1}{N_1} \times 100$$

where
N_1 = value of the variable at time 1
N_2 = value of the variable at time 2

For example, assume that 5,000 individuals were homeless in the first year (time 1) and 4,325 the second year (time 2). The calculations to determine the percentage change would be as follows:

$$
\begin{aligned}
\text{Percentage change} &= \frac{4,325 - 5,000}{5,000} \times 100 \\
&= \frac{-675}{5,000} \times 100 \\
&= -0.135 \times 100 \\
&= -13.5\% \text{ change or } 13.5\% \text{ decrease}
\end{aligned}
$$

The percentage change can be positive or negative. If the number of homeless in the second year was 5,075, the percentage change would be 1.5%, indicating a 1.5 percentage increase in the homeless population.

CHARACTERISTICS OF A DISTRIBUTION

While a frequency distribution includes useful information, you may want a simple statistic to summarize its content. Measures of central tendency reduce the distribution to a value that represents a typical case, the center of the distribution, or both. Measures of central tendency give an incomplete picture of how typical the typical case is. Focusing only on a typical case may be misleading, since cases may be widely different from one another. Measures of variability fill in this gap and add to your knowledge about the distribution. Measures of variability show how representative the typical case is by giving you information on how spread out or dispersed all of the cases are and how far they are from a central point. Several statistics are used to measure central tendency and variation. The choice of a statistic depends on the level of measurement of the variable and what information you will find most valuable. Note that measures of central tendency and variability are the same for interval and ratio scales, so to simplify our discussion we will refer only to ratio scales.

Measures of Central Tendency

Measures of central tendency indicate the value that is representative, most typical, or central in the distribution. The most common measures are the mode, median, and arithmetic mean.

Mode: The simplest summary of a variable's frequency distribution is to indicate which category or value is the most common. The *mode* is the value or category of a variable that occurs most often. In a frequency distribution it is the value with the highest frequency. Table 5.3 shows a distribution of a variable to the Happy Housing Center, Reason for Requesting Services. The mode is Needs Help Finding An Apartment. More clients came for that reason than for any other. Of course that reason has the highest relative frequency as well.

The mode can be determined for all measurement scales: nominal, ordinal, interval and ratio. If two values have the same frequency, the variable has two modes and is said to be *bi-modal*. A common mistake is to confuse the frequency of the modal category with the mode. For instance, the mode for the Reason for Requesting Services in Table 5.4 is Needs Help Finding an

TABLE 5.4		
Frequency and Percentage Distributions for Reason For Requesting Services From Housing Center.		
Reason for Requesting Services	Number of Clients ($N = 50$)	Percent
Has Been Evicted	7	14
Needs Help Finding an Apartment	13	26
Rent Has Been Increased	8	16
Lost Job	12	24
Rental House Needs Repairs	10	20

Apartment. It is not 13. For nominal and many ordinal variables, the category that occurs most often will have a name; for ratio variables, the value of the mode will be a number.

Median: The *median* is the value or category of the case that is in the middle of a distribution in which the cases have been ordered along a continuum. It is the value of the case that divides the distribution in two; one-half of the cases have values less than the median and one-half of them have values greater than the median. The median requires that variables be measured at the ordinal or ratio level. To find the median, as mentioned, you must order the case values along a continuum. It makes no sense to find the middle case if the cases have not been ordered according to their values on the variable of interest.

To find the median, you locate the middle case in a distribution. If the number of cases is odd, then the median is the value of a specific case. If the number of cases is even, then the median is estimated as the value halfway between two cases. For example, with 11 cases, the median is the value of the 6th case. If there are 12 cases the median is halfway between the value of case number 6 and case number 7. The formula for finding the middle case is $(N + 1)/2$.

Two examples follow. Table 5.5 shows the distribution of an ordinal variable. The table reports 11 clients' ratings of a Happy Housing Center transportation program.

The middle case is the case number 6, which is determined by dividing the number of cases plus 1 by 2: $(N + 1)/2$. This case has the value Neither Good Nor Poor. The median for this variable, then, is Neither Good Nor Poor. One-half of the cases rated the transportation program as Neither Good Nor Poor or better, and one-half rated it as Neither Good Nor Poor or worse. Table 5.5 has two values with the largest number of ratings, Neither Good Nor Poor, Very Good, etc. This is an example of a bi-model distribution with both Neither Good Nor Poor and Very Poor as modes.

TABLE 5.5

Ranking of Transportation Support

Client	Rating	Code in Excel File
B	Very Good	5
E	Very Good	5
H	Good	4
G	Neither Good Nor Poor	3
C	Neither Good Nor Poor	3
J	Neither Good Nor Poor	3
D	Poor	2
F	Poor	2
A	Very Poor	1
I	Very Poor	1
K	Very Poor	1

Table 5.6 shows the frequency distribution for a ratio variable, the number of months spent in temporary shelter. The data have been arranged according to increasing values. Since the number of cases, 16, is even, the middle case is case 8.5; the middle of the distribution is between case number 8 and case number 9. Therefore, the value of the median is 12.5.

▶ **TABLE 5.6**
Number of Months in Temporary Shelter
Number of Months   7, 7, 7, 9, 10, 11, 11, 12, 13, 14, 15, 15, 16, 16, 17, 18

The mode for Table 5.6 is 7 months. If we only use the mode, the additional information provided by the median is lost. The median is the preferred measure of central tendency for ordinal variables and for ratio variables that have a few extreme values. A distribution with a few extreme values is said to be skewed. Since the median is little affected by extreme numerical values, it gives a more accurate picture of central tendency than the arithmetic mean, which is discussed next.

Arithmetic Mean: A third measure of central tendency is the arithmetic average or *arithmetic mean*. When you first learned this measure in school you simply called it "the average." The mean is the appropriate measure of central tendency for variables measured at the ratio level. To calculate the mean, add the value of the variable for each case and divide it by the number of cases. The formula for the arithmetic mean is

$$\text{Arithmetic mean} = \text{sum } x_i/N.$$

where sum x_i indicates the sum of all the cases and N indicates the number of cases. The data for Table 5.6 has a mean of 12.38 (value of all cases/number of cases = 198/16).

If you are measuring a variable representing resources, beware that the mean can seriously misrepresent the data. Such resources include income, value of stocks, bonds or real estate, or size of land holdings. If a data set in a small community included a person with an income well over $1,000,000 the mean may greatly overestimate the typical income of the remaining less-fortunate members of the community. With measures of resources and other skewed data the median is the appropriate measure of central tendency.

Measures of Variation and Dispersion

You may find that policy makers and the general public are uneasy with just one value to summarize a variable. Through education, experience, or skepticism they may suspect that a median or mean doesn't tell the whole story. So in addition to measures of central tendency you want to see how far the values spread out from the central part of the distribution. *Measures of dispersion* describe the similarity of the data. Relatively smaller values of the measure of dispersion for a variable imply more uniformity, whereas relatively larger values imply more diversity or variation. Some measures indicate only the difference between two observations in an ordered set of values. Other measures consider all observations in a distribution.

Consider the average number of months that clients in two cities lived in temporary shelter. If you calculate statistics using the data in Table 5.7 you will find the average stays of clients in temporary housing in City A and City B to be very similar. In analyzing the data, however, you can see that individual stays are very different in these two cities. Measures of dispersion will provide more complete information and avoid implying that the lengths of stays in the two cities are virtually identical. Maximum variation for ratio variables occurs when all cases are equally divided between two extreme values. Maximum variation for nominal and ordinal variables occurs when cases are evenly distributed across all categories. The more the cases are clustered in one category, the less the variation. If all cases were in one category, then variation would be zero. Common measures of dispersion are the range, the inter-quartile range, and standard deviation.

TABLE 5.7

Number of Months Clients Spent in Temporary Shelter by City

City A	7, 7, 7, 9, 10, 11, 11, 12, 13, 14, 15, 15, 16, 16, 17, 18
City B	3, 3, 4, 5, 7, 8, 10, 11, 12, 13, 14, 16, 17, 21, 22, 24

Range: The simplest measure of dispersion is the *range*, which is the difference between the highest value and the lowest value in a distribution. For City A, the range is 18 − 7, or 11 months. For City B the range is larger, 24 − 3 or 21 months. Whether you are reporting variations among regions or respondents, users can easily interpret the range; for some purposes you may find that reporting the highest and lowest values is clearer or more effective. You could report, for example, that the length of stay for City A was from 7 to 18 months and for City B from 3 to 24 months.

Inter-quartile Range: A range can be unduly affected by one value that is well above or well below the other values in the distribution. The inter-quartile range (IQr) identifies the values for the middle 50 percent of cases and is not affected by extreme values. The first quartile is that value below which 25 percent of the cases are found. The third quartile is that value below which 75 percent of the cases are found. The second quartile, of course, is the median. In determining the inter-quartile range, the lowest 25 percent of the observations and the highest 25 percent are omitted (see Figure 5.5). As with the median, determining the quartiles requires that you order the cases according to their values.

Let's look at City A in Table 5.7. The range of the middle 50 percent of values includes the eight values between 10 and 15. To find the first quartile drop the lowest 25 percent of the 16 cases, that is, the lowest 4 cases. To find the third quartile drop the upper 25 percent of the 16 cases, that is, the 4 cases with the highest values. For this distribution the inter-quartile range would be 15 − 10 = 5. The spread of the middle 50 percent of the cases is 5 months. (Technically the value of the cutoff for the first quartile would be 9.25—25 percent of the way between 9 and 10—and for the third quartile it would be 75 percent of the way between 15 and 16. However, quartiles are often estimated as shown here.)

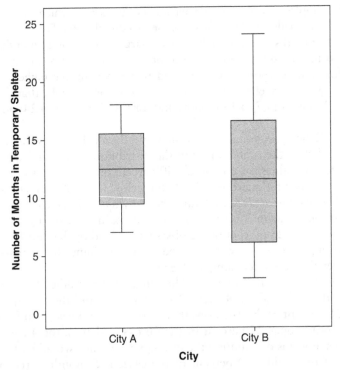

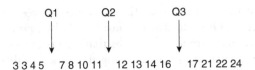

FIGURE 5.5
Identifying Quartiles

The IQr for City B is $16 - 7 = 9$ months. (Its third quartile is 75 percent of the way between 11 and 13, but presenting a rough estimate of the IQr is acceptable.) The middle 50 percent of the cases are spread over a range of 9 months.

By itself the inter-quartile range may have little value. It is most useful to compare distributions from one time to another or between two or more groups. For example, policy makers can examine the change in the annual rates of infant mortality for all counties in one state over time. Or they can compare how those rates varied in any of the 50 states at one time. In both cases the inter-quartile range will provide more information than a measure of central tendency and the range.

Audiences who are less comfortable with statistics or those who are visually oriented seem to become engaged with box plots. A *box plot* places either the mean or the median in a box enclosed by the IQR and a line extends out to the range. Figure 5.6 presents the city data as box plots.

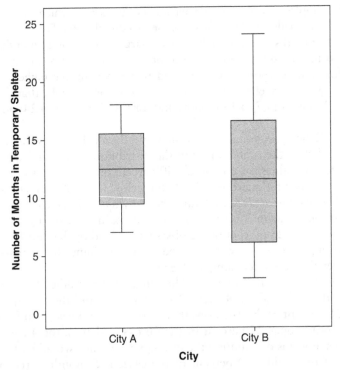

FIGURE 5.6
Boxplots of Number of Months in Temporary Shelter

Standard Deviation: If you have taken a basic statistics course, you are familiar with the standard deviation. It is the most common measure of the variation in a distribution. The standard deviation is calculated by

- subtracting the mean from each individual value. These deviation values show how much the value of each case deviates from the mean of the distribution;
- squaring each deviation value;
- adding the squared deviations;
- dividing the sum of squared deviations by the number of cases (N);
- taking the square root of the divided sum of squared deviations.

Table 5.8 uses the City A data from Table 5.7 to calculate the standard deviation. The mean stay in temporary shelter was 12.38 months.

TABLE 5.8

Calculating Standard Deviation

$x - \bar{x}$	$(x - \bar{x})^2$
$7 - 12.38 = -5.38$	28.9
$7 - 12.38 = -5.38$	28.9
$7 - 12.38 = -5.38$	28.9
$9 - 12.38 = -3.38$	11.4
$10 - 12.38 = -2.38$	5.7
$11 - 12.38 = -1.38$	1.9
$11 - 12.38 = -1.38$	1.9
$12 - 12.38 = -0.38$	.1
$13 - 12.38 = 0.62$	.4
$14 - 12.38 = 1.62$	2.6
$15 - 12.38 = 2.62$	6.9
$15 - 12.38 = 2.62$	6.9
$16 - 12.38 = 3.62$	13.1
$16 - 12.38 = 3.62$	13.1
$17 - 12.38 = 4.62$	21.3
$18 - 12.38 = 5.62$	31.6
Total	203.6

Standard deviation $= \sqrt{(203.6/16)} = 3.57$

A major use of the standard deviation is to estimate the variation in a population. It is important in statistics, because you can use data from a sample to estimate the distribution of a population. To estimate the variation in the population you have to assume (or verify) that the frequency distribution is reasonably described by the normal curve. A *normal curve* is bell-shaped and symmetrical so that the mode, median, and mean all have the same value. When working with a sample you will want to put $n - 1$ in the denominator to calculate the standard deviation; $n - 1$ gives a less biased estimate of the population variance. For example, assume that the data represent a random sample of residents staying in temporary shelter in City A. We would put 15 in the denominator, and the recalculated standard deviation would be 3.69. Using the mean and the standard deviation you may estimate the following from the data in Table 5.7:

1. 50 percent of the observations are above the mean and 50 percent are below it.
2. 34.1 percent of the observations are one standard deviation above the mean and 34.1 percent are below it. So you would estimate that 68.2 percent of the people who stay at a temporary shelter in City A stay there between 8.69 and 16.1 months.
3. 47.7 percent of the observations are two standard deviations above the mean and 47.7 percent are below it, or roughly 95 percent of the observations are within two standard deviations of the mean. So you would estimate that 95 percent of the people who stay at a temporary shelter in City A stay there between 5 and 19.8 months.
4. 49.8 percent of the observations are three standard deviations above the mean and 49.8 percent are below it, or over 99 percent of the observations are within three standard deviations of the mean. So you would estimate that 99 percent of the people who stay at a temporary shelter in City A stay there between 1.31 and 26 months.

CONCLUDING OBSERVATIONS

As you organize data you begin a learning process. If you began with a logical model you can use the tools discussed in this chapter to describe a program's inputs, activities, outcomes, and outputs. This information may motivate you and others to identify parts of the program that are working better than expected and parts that are under-performing. You may begin to wonder why the various components are the way they are and what needs to be done to strengthen some components and to mirror the success of others.

You may be stimulated to take action and monitor whether it had an impact.

Frequencies, rates, percentage changes, and measures of central tendency are tools that you will apply time and time again whether you are simply describing a data set or conducting more elaborate studies to see what works and under what conditions. The graphs may be less useful in conducting your own analysis, but they are an essential tool in communicating your findings to others.

RECOMMENDED RESOURCES

Best, J. *Damned Lies and Statistics: Untangling Numbers from the Media, Politicians, and Activists* (Los Angeles, CA: University of California Press, 2001).

Rumsey, D. *Statistics for Dummies* (Hoboken, NJ: Wiley Publishing, Inc., 2009).

O'Sullivan, E., G. Rassel, and M. Berner. *Research Methods for Public Administrators*, Fifth edition (New York: Pearson/Longman, 2008). Chapter 11.

Salkind, Neil. *Statistics for People Who (Think They) Hate Statistics*, Second edition/Excel 2007 (Thousand Oaks, CA: Sage Publications, Inc., 2010).

CHAPTER 5 EXERCISES

There are two separate exercises for Chapter 5. Each exercise develops your competence in interpreting and applying measurement concepts.

- Exercise 5.1 Fresh Start Center focuses on a database for a community job training program. The exercise presents a partial database. You are asked to use graphs and quantitative measures to describe the variables and decide on effective strategies to present the information.
- Exercise 5.2 Purple Flower Neighborhood Association asks you to apply basic skills to examine some crime data.

EXERCISE 5.1 Fresh Start Center

Scenario

Fresh Start Center is a community partnership of a community college, county government, local businesses, and nonprofits to offer job training to unemployed workers. It offers culinary training (training to work in restaurants or with caterers), automotive repairs, and carpentry. You have been asked to help the center develop a plan for monitoring its performance. Table 5.9 includes 25 cases from the first round of data collection. The data report the number of months of employment prior to entering training and the employment status of June 30 graduates as of August 1.

Section A: Getting Started

You may enter the Table 5.9 data in Excel to carry out the following tasks.

1. Create a frequency and relative frequency distribution for each variable. As appropriate, combine categories.
2. Create pie charts for perceived technical skills and employment status.
3. Create bar charts for perceived technical skills and employment status.
4. Create a histogram for months unemployed.
5. For each variable indicate.
    a. its level of measurement (nominal, ordinal, interval/ratio);
    b. the value of the mode, median, and mean, as appropriate.
    (Directions are at the end of exercise.)
6. Create a box plot for months unemployed and perceived technical skills. What do they suggest about the variation (dispersion) of the two variables?
7. Use the data to write a paragraph for the center's annual report. Would you include graphics? Which one(s)?
8. Which variables should the agency track? Explain your choices.

TABLE 5.9

Fresh Start June 30 Graduates

Id	Months Unemployed	Education	Skill Rating	Training Program	Status on August 1
1	1	HS Grad	14	Culinary	FT Work
2	8	> HS	15	Automotive	Unemployed
3	7	> HS	14	Automotive	PT Work
4	11	> HS	14	Automotive	FT Work
5	1	HS Grad	12	Culinary	PT Work
6	8	> HS	13	Automotive	Unemployed
7	3	College Grad	12	Carpentry	Unemployed
8	11	College Grad	13	Automotive	Unemployed
9	5	< HS	13	Carpentry	FT Work
10	1	> HS	11	Carpentry	FT Work
11	4	HS Grad	14	Carpentry	PT Work
12	4	> HS	12	Carpentry	FT Work
13	9	HS Grad	14	Automotive	Unemployed
14	6	> HS	13	Automotive	FT Work
15	1	> HS	16	Culinary	Unemployed
16	1	>HS	13	Culinary	Unemployed
17	6	> HS	13	Carpentry	FT Work
18	6	> HS	10	Carpentry	Unemployed
19	5	> HS	12	Carpentry	Unemployed
20	4	< HS	12	Automotive	FT Work
21	6	< HS	14	Automotive	Unemployed
22	8	HS Grad	14	Automotive	Unemployed
23	8	HS Grad	12	Automotive	FT Work
24	1	> HS	15	Culinary	Unemployed
25	1	> HS	11	Culinary	FT Work

Skill Rating: Just prior to graduation the trainees rated their agreement with the following statements
(4 = Strongly agree, 3 = Agree, 2 = Disagree, 1 = Strongly disagree)
I can choose the right tool or equipment for the job at hand.
I know how to take care of the equipment
I have the skills I need to do my work
I have the training needed to do my work

Section B: Small Group or Class Discussion

1. One of the challenges is to sort through a data set and decide what is important. Although Table 5.9. represents only a partial data set, based on your
analysis consider each variable. Which of the following would you use to
present the data: a frequency distribution, a graph (which type), a measure of
central tendency, a measure of dispersion? Would you categorize the values of
Months Unemployed and Skill Rating? Why or why not? If you would categorize them, how would you organize the data?

2. Which variables would you suggest that the center collect regularly? For each variable consider how they should report the data (as a frequency, percentage, median or mean, or something else)? Explain the thinking behind your recommendations.

EXERCISE 5.2 Purple Flower Neighborhood Association

Scenario

The Purple Flower neighborhood Association's mission is to "improve neighborhood safety, beautification, and education and to represent community interests to the city of Lilac Fields (population 94,700)." Each year association committees recommend objectives for the next year. The safety committee examined the problems of burglary and vandalism.

Section A: Getting Started

1. Based on newspaper reports of a rash of burglaries, community residents contend that the neighboring towns of Maple Leaf (population 51,600) and Elm Hill (population 12,700) are safer. Last year Lilac Fields had 705 burglaries; Maple Leaf had 215; Elm Hill had 72).

   Compare the burglary rates in these three towns. Two years ago Lilac Fields had 549 burglaries. What was the percentage increase?

2. The committee gathered data on reports of vandalism in the neighborhood. Table 5.10 displays these data.

TABLE 5.10

Reported Incidents of Vandalism in Purple Flower by Year

1994	53
1995	52
1996	51
1997	53
1998	56
1999	57
2000	56
2001	56
2002	57
2003	58
2004	59
2005	58
2006	57
2007	59
2008	60
2009	61

a. Draw a time series graph of these data.

b. Identify the variations in the time series.

c. If a new development opened in Purple Flower in 1999 should the committee track the number of incidences or something else? Explain.

d. To calculate the vandalism rate the committee debates using (i) the number of residences, (ii) the population, or (iii) the number of residents between 10 and 19 years of age for the denominator in the rate formula. Which would you suggest? Why?

CALCULATING STATISTICS WITH EXCEL

After you input the data you can access routines to calculate the data using the Function Wizard. Depending on your version of Excel, you may have to hunt for the Function Wizard; on some versions you may find function (fx) under "Insert."

Click on fx; a dialogue box will appear.

Click on arrow by "Search for a Function" or "Select a category"; a drop down menu listing Financial, Statistical, Math & Trig, and so on will appear.

Click on Statistical; a drop down menu listing statistical routines, such as AVERAGE, MEDIAN, and STDEV will appear.

Click on the function you want; another dialogue box will appear.

Enter the range of cells containing the data to analyze. For example, if Months unemployed data are in column B, rows 2–26, put B2:B26 in the first line in the dialogue box. You should get 5.04 for the value of the arithmetic mean of the number months unemployed. ■

Citizen Surveys: Their Design and Analysis

Chapters 6 through 10 cover the skills you need to conduct and analyze survey data. Managers who fail to use the skills covered in these chapters miss tapping into important information on how well their organizations are doing. You may feel that surveys are overused; we agree. The skills covered in these chapters are useful not only for surveys but also for other activities where you need to write questions or decide whom to contact. Similarly, the methods of analysis covered in Chapters 8 through 10 are applicable in other types of studies. Chapters 8 and 9 build on Chapter 5. Chapter 8 discusses strategies for describing a data set and looking at relationships among variables. If you have data from a probability sample you may be able to generalize your findings beyond your sample—this is the subject of Chapter 9. A word of warning—parts of Chapters 8 and 9 may be difficult if you have not had a statistics course; your instructor may be able to supplement our presentation. Our experience has shown that by working through this material even beginners can understand it. Chapter 10 guides you through the process of collecting and analyzing data through interviews and asking open-ended questions.

Selecting and Contacting Subjects

A s you think about surveying clients, donors, or the general public you may start by figuring out whom to contact. Contacting everyone is seldom practical or necessary. You may use too many resources and take up too much time. Instead you will want to contact a sample of clients, donors, or residents. This chapter covers the principles guiding the selection of a sample. Knowing how to design a sample is only the beginning. You need to identify individual sample members, contact them, and encourage them to respond. Technology is rapidly changing strategies for contacting sample members and encouraging them to respond. Survey organizations are studying the impact of changing lifestyles and new communications technologies. In this chapter we focus on the basic sampling and data collection strategies. At the end of the chapter we refer you to how-to books and other sources to fill in the gaps. Our approach keeps us from overwhelming you with information that may have a short shelf life.

THE PRINCIPLES OF SAMPLE DESIGN

Sampling is an economical and effective way to learn about individuals and organizations. This is true if you seek information from individuals, case records, agency representatives, or computerized datasets. Even with a relatively small group, sampling enables you to gather information relatively quickly. Depending on how you select the sample you may be able to use the findings to generalize beyond the sampled individuals, organizations, or other units. Because sampling is a jargon-rich topic, we begin by defining some key terms.

Sampling Terminology

First, you want to recognize the difference between a population and a sample. The *population* is the total set of units that you are interested in. A population is often composed of individuals, but it may consist of other units such as organizations, households, records, or computers. A *sample* is a subset of the population. Closely connected to the concepts of population and sample are the terms parameter and statistics. A *parameter* is a characteristic of the entire population. A *statistic* is a characteristic of a sample. The percentage of citizens in a city favoring a fund to finance affordable housing is a parameter. The percentage of sampled residents who favor the fund is a statistic. Depending on how you select the sample you may be able to use the statistic and the sampling error to estimate the parameter. The *sampling error,* which is discussed in more detail in Chapter 9, is the difference between the parameter and the statistic and is used to estimate the parameter.

One of your first steps is to precisely define the population represented by the sample, such as "all adults over the age of 21 living in Clark County on July 1." After you define your population your next step is to find a sampling frame. A *sampling frame* is a list of names of individuals or organizations in your population. A sampling frame should contain all members of your population. Complete lists seldom exist. So, in reality the sampling frame is the list of potential sample members. Sampling frames are not perfect. They may aggregate units, list units more than once, omit units, or include units that are not part of the population. A list of building addresses, for example, may include apartment buildings, but not individual units. If you plan to sample individual residences, this sampling frame would be inadequate.

The *unit of analysis* indicates what the data represent. You may think of the unit of analysis as what constitutes a case when you enter data on a spreadsheet. If you collect data from housing agencies, the sampling unit is housing agency. If you report the number of employees and the size of the budget the unit of analysis is housing agency. If you collect data on individual clients from the sampled agencies, then "client" would be the unit of analysis. Figure 6.1 illustrates the components of a sample.

A *sample design* describes the strategy for selecting the sample members from the population. Designs are classified as probability and nonprobability. With a probability sample, each unit of the population has a known, nonzero chance of being in the sample. Probability samples use randomization to avoid selection

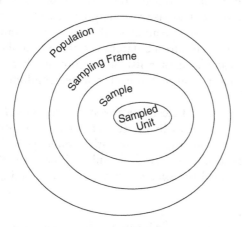

FIGURE 6.1
From Population to Sampled Unit

AN APPLICATION OF SAMPLING TERMINOLOGY

- *Population.* All participants who have completed the Clark County Job Training Program within the past 2 years
- *Parameter.* Average starting salaries of all graduates of the Clark County Job Training Program within the past 2 years
- *Sampling frame.* All (8) graduation lists from the past 2 years

- *Sampling design.* Probability sample
- *Sample.* 120 graduates randomly selected from sampling frame
- *Unit of analysis.* Individual graduates
- *Statistic.* The average starting salary of the 120 graduates was $25,000 ∎

biases and use statistical theory to estimate parameters. You cannot estimate parameters with nonprobability samples. Both types of designs have value but for different purposes and in different situations.

Probability Samples

With a probability sample, you can calculate the chance that a member of the population has of being sampled. We discuss four common probability sampling designs: simple random sampling, systematic sampling, stratified sampling, and cluster sampling. These designs are often used together. Stratified and cluster samples also use simple random sampling or systematic sampling to select subjects. Multistage sampling combines cluster sampling with other designs.

Simple random sampling requires that each unit in the population has an equal probability of being in the sample. Drawing names from a hat is the prototypical example of a simple random sample. An analogous strategy is used to draw lottery numbers. You do not need to put names in a hat or marked balls in a tumbler. Rather you can use technology to help you draw a random sample. If the cases are contained in an electronic database you may use a computer program to select your sample. These programs use an algorithm to create a random sample. If the cases are not stored in an electronic database you can number the cases. In this method, using your calculator or a list of random numbers, you select a set of random numbers. Finally, you match each selected random number to the case with the same number.

If the cases are not stored electronically *systematic sampling* is an acceptable alternative. It is easier and normally produces results comparable to those of a simple random sample.[1] To construct a systematic sample, you divide the number of units in the sampling frame (N) by the number desired for the sample (n). The resulting number is called the *skip interval* (k). If a sampling frame consists of

---

[1]Systematic sampling does not result in a truly random sample. Although each unit in the population has the same chance of being selected, adjacent units within the skip interval will not be selected. Systematic sampling is used widely, however, and works well.

50,000 units and a sample of 1,000 is desired, the skip interval equals 50 (50,000 divided by 1,000). You select a random number, go to the sampling frame, and use the random number to select the first case. You then pick every *k*th unit for the sample. In our example every 50th case would be chosen. If the random number 45 was selected the cases 45, 95, 145, and so on would be in the sample. With systematic sampling, treat the list as circular, so the last listed unit is followed by the first. You should go through the entire sampling frame at least once.

Systematic sampling has one potentially serious problem, that of *periodicity*. If the items are listed in a pattern and the skip interval coincides with the pattern, the sample will be biased. Consider what could happen in sampling the daily activity logs of a 911 call center. If the skip interval was 7 the activity logs in the sample would all be for the same day, that is, all Mondays or all Tuesdays, and so forth. A skip interval of any multiple of 7 would have the same result. Periodicity is relatively rare. If you notice something strange about your sample, for example, that it is all female, includes only top administrators, or has only corner houses, check for periodicity. An easy fix that often works is to double the size of your skip interval ($k \times 2$). Go through the sampling frame once to get half your sample, then choose another random starting point and go through it a second time.

Systematic sampling may be the only feasible way to get a probability sample of a population of unknown size, such as people attending a community festival. You estimate the population size, that is, the number of attendees, determine a skip interval, and pick a random beginning point. If you plan to sample every 20th person and start with the 6th person, the 6th person to arrive (or depart) would be selected, as would the 26th person, the 46th person, and so on, until the end of the sampling period.

Stratified sampling ensures that a sample adequately represents selected groups in the population. You should consider using stratified sampling if you plan to compare groups, if you need to focus on a group that is a small proportion of your population, or if your sampling frame is already divided by groups. First, you divide or classify the population into strata, or groups, on some characteristic such as gender, age, or institutional affiliation. Every member of the population should be in one and only one stratum. Use either simple random sampling or systematic sampling to draw samples from each stratum.

Stratified samples may be proportional or disproportional. In *proportional stratified samples* the same sampling fraction is applied to each stratum. The sample size for each stratum is proportional to its size in the population. If, for example, you wanted to compare three groups of employees—professional staff, technical staff, and clerical staff—you would designate each group as a stratum. You would select your samples by taking the same percentage of members from each stratum, say 10 percent of the professional staff, 10 percent of the technical staff, and 10 percent of the clerical staff. The resulting sample would consist of three strata, each equal in size to the stratum's proportion of the total population. Example 6.1 illustrates a proportional stratified sample.

The percentage of members selected for the sample need not be the same for each stratum. In *disproportional stratified sampling*, the same sampling fraction is not applied to all strata. Disproportional sampling is useful when a characteristic of interest occurs infrequently in the population, making it unlikely that a simple

EXAMPLE 6.1 AN APPLICATION OF PROPORTIONAL STRATIFIED SAMPLING

- *Problem.* The director of volunteer recruitment for the state office of a large nonprofit organization wanted information about volunteer applicants to the organization over the previous 4 years.
- *Population.* All applications to the volunteer service of the nonprofit organization for each of the past 4 years.

- *Sampling frame.* Agency's electronic file of applications(organized by application date).
- *Sampling design.* Proportional stratified sample with year of application as strata. A computer program selects a random sample of applications for each year, using the same sampling fraction for each group (Table 6.1). ∎

TABLE 6.1

Proportional Stratified Sampling

Strata	Number of Applications	Sampling Fraction (%)	Number in Sample
Year 1	350	15	53
Year 2	275	15	41
Year 3	250	15	38
Year 4	230	15	35
Total	1,150		167

random or a proportional sample will contain enough members with the characteristic to allow full analysis. For example, a mentoring program may want to compare recruitment and retention of Spanish-speaking volunteers with that of non–Spanish speakers. You will want a sample large enough so that you can analyze and compare the two groups. If 5 percent of its volunteers are Spanish speakers, a random sample of 200 volunteers should have about 10 Spanish speakers—too few cases for analysis. Therefore, you might survey a larger percentage of Spanish-speaking volunteers and a smaller percentage of other volunteers; you could decide that half the sample should be Spanish speakers and an equal number of non-Spanish speakers, thus over-representing Spanish speakers. A higher percentage of Spanish speakers will have been sampled and non-Spanish speakers under-represented.

Each strata in a disproportional stratified sample constitutes a separate sample. To conduct your analysis you must keep the samples separate as you compare the Spanish-speaking volunteers with non–Spanish speakers. To determine characteristics of all sampled volunteers you must weigh the samples. First, determine each stratum's sample size for a proportional sample. To calculate the weight for each stratum, divide the size of the proportional sample by the size of the disproportional sample.

Strata	Population	Sample	If proportional	Weight
Spanish Speakers	200	100	$(200 \times 0.05) = 10$	$10/100 = 0.1$
Non–Spanish Speakers	4,000	100	$(200 \times 0.95) = 190$	$190/100 = 1.9$

When you combine the samples, each Spanish speaker's responses would be multiplied or weighted by 0.1 and each response by non–Spanish speakers would be multiplied by 1.9.

Cluster and multistage sampling take advantage of the fact that members of a population can be located within groups or clusters, such as cities and counties. Cluster sampling is useful if a sampling frame does not exist or is impractical to use. For cluster and multistage samples, we randomly select clusters and then units in the selected clusters. For example, for a study of Boys and Girls Clubs we might first select a sample of counties. Either our final sample would consist of all the Boys and Girls Clubs in the sampled counties or we would select a sample of clubs from these counties. Unlike stratified samples, which sample members from each stratum, cluster samples sample only members from the selected clusters.

Multistage sampling is a variant of cluster sampling. It proceeds in stages as you sample units dispersed over a large geographic area. The following example shows how you could design a sample to survey residents in a state's long-term care facilities:

- *Stage One.* Draw a probability sample of counties, that is, you choose large clusters containing smaller clusters.
- *Stage Two.* From your sample of counties choose a probability sample of incorporated areas.
- *Stage Three.* For your sampled incorporated areas obtain lists of long-term care facilities located in them.
- *Stage Four.* Select a sampling strategy
  - *Alternative 1.* Select all residents in the facilities identified at Stage 3 or
  - *Alternative 2.* Select a sample of long-term facilities and then select all residents in the selected facilities or
  - *Alternative 3.* Select a sample of long-term facilities and select a sample of residents in the selected facilities.

The units selected at each stage are called *sampling units*. The sampling unit may not be the same as the unit of analysis. Different sampling units are selected at each stage. In this example, the unit of analysis is residents in long-term care facility. However, different sampling units were selected at each stage of the process. Note that you can incorporate different sampling strategies into this design. For example, you may use stratified sampling at Stage One to make sure that you have counties for each of the state's major regions, such as upstate or downstate, or the mountains and the beaches. You may use simple random sampling to choose specific residents for your sample.

Cluster sampling is recommended if your population is distributed over a large geographic area. Without the ability to limit the sample to discrete areas the costs and logistics would make probability sampling difficult, if not impossible. If site visits are needed to collect data, visiting a sample spread thinly over a wide area will be extremely expensive. Cluster sampling can help reduce this cost. Cluster sampling also helps compensate for the lack of a sampling frame. Combining cluster and multistage sampling requires an investigator to develop a sampling frame for just the last stages of the process.

Although cluster and multistage sampling methods reduce travel time and costs, they require a larger sample than other methods for the same level of accuracy. In the multistage process, probability samples are selected at each stage. Each time a sample is selected there is some sampling error; thus, the overall sampling error is likely to be larger than if a random sample of the same size were drawn.

Nonprobability Samples

Nonprobability samples cannot produce estimates with mathematical precision, because the sample members do not have a known chance of being selected. Nonprobability sampling is useful when designing a probability sampling is not possible, studying small or hard-to-reach populations, doing an exploratory or preliminary study, or conducting in-depth interviews or focus groups. In the following paragraphs we describe four common nonprobability designs: availability sampling, quota sampling, purposive sampling, and snowball sampling.

Availability Sampling: Availability or convenience sampling is done when cases are selected because they are easily accessed. Subjects may be selected because they can be contacted with little effort. For example, if an emergency shelter wanted to interview clients about the service availability and quality, everyone at the shelter on a given day might be invited to participate.

The obvious flaw of availability sampling is that it may exclude cases that represent the target population, and the findings cannot be generalized. The findings do not describe the knowledge, attitudes, beliefs, or behaviors of others in the target population. Consider the study at the emergency shelter. You may avoid approaching certain people. For instance, individuals who are sleeping or nodding off may not be asked to participate. These individuals may be clients of a methadone treatment program (methadone causes severe drowsiness), and their observations about drug treatment services may go unheard. If you conduct the study in the summer, the population may be very different from the winter population. The day staff may have a reputation that affects who stays around the shelter during the day.

Quota Sampling: Quota sampling, often used in market research, is less common in social science studies. Quota samples attempt to overcome the major limitation of availability samples by defining the percentage of members to be sampled from specified groups. An agency considering opening a child care center in an ethnically diverse community may want to survey local families. You could design a quota sample to ensure that you get input from families of all ethnicities. If the community is 25% Asian, 60% White, 10% African American, and 5% other, a sample of 200 should have 50 Asians, 120 Whites, 20 African Americans, and 10 from other groups. If the selection of sample members is based on convenience and judgment, they may also have characteristics that make them more approachable. This adds a potential bias to the sample.

While your sample may be representative of one variable, it is not necessarily representative of other variables. For example, your sample of families based on their ethnicity probably isn't representative of age or income. As part of designing a quota sample you need to determine the most meaningful variable and its variation in the population. It may be easy to find information on some variables (e.g., race, gender, age) but more difficult to find out about other characteristics such as incidence of substance abuse or child-rearing practices.

Purposive Sampling: Purposive sampling selects cases based on specialized knowledge, distinct experience, or unique position. You might use a purposive sample to study very successful mentoring programs, people who underwent innovative medical treatments, or women governors. You might select cases to capture maximum variation of a phenomenon. The selected cases may provide rich, detailed information about the phenomenon of interest. Some researchers have argued that we can learn more from very successful programs and limit their samples to such programs. A variation of this theme is to compare a set of very successful programs with unsuccessful ones.

If you want to conduct in-depth interviews or focus groups you may want to select respondents deliberately. You will want to carefully select whom to interview. You will want to interview individuals who can provide insight about the phenomenon being studied, are willing to talk, and have diverse perspectives. Such individuals are referred to as key informants. Good informants will have given the subject matter some thought and can express their thoughts, feelings, and opinions.

Snowball Sampling: Snowball sampling starts by finding one member of the population of interest, speaking to him or her, and asking that person to identify other members of the population. You can use it to identify organizations or individuals. This process is continued until the desired sample has been identified. The number of members in the sample "snowballs." Snowball sampling is most often used to contact individuals or groups that are hard to reach, difficult to identify, or tightly interconnected. For example, if you wanted to interview sex workers, you might identify a sex worker who is willing to talk to you and to identify other sex workers. In addition to identifying other sex workers, an informant may also vouch for your credibility.

A concern when using snowball sampling is that the individuals who are referred have not consented to being identified. This may not be a problem for some populations, but it is for others. Consequently, for this type of sampling, perhaps more than for others, you should remind any informant of the need to protect people's privacy and have the informant seek permission before sharing names and contact information.

SAMPLE SIZE
For Qualitative Studies

When conducting qualitative research including interviews and focus groups, determining how many cases to include in the sample may be difficult. Completeness and saturation serve as a guide for knowing when to stop selecting participants.[2]

---

[2]H. Rubin and I. Rubin, *Qualitative Interviewing: The Art of Hearing Data* (Thousand Oaks: Sage, 1995).

Completeness suggests that the subjects have given you a clear, well-defined perspective of the theme or issue. *Saturation* means that you are confident that little new information will be learned from more interviews or focus groups. Once you find that your newest subjects are sharing the same ideas, themes, and perspectives as previous participants, you can stop.

For Probability Samples

The appropriate sample size for quantitative surveys is based on several factors. The first is how much sampling error you are willing to accept, since accuracy is important in determining sample size. Greater accuracy usually can be obtained by taking larger samples. The confidence you wish to have in the results and the variability within your target population also play a role in determining sample size. Both the desired degree of confidence and the population variability are related to accuracy. We discuss these terms in more detail and their relation to sample size in Chapter 9.

People unfamiliar with sampling theory often assume that a major factor determining sample size is the size of the entire population. That is, they may believe that representing a population of 500,000 requires a correspondingly larger sample than representing a population of 20,000. Generally, larger samples will yield better estimates of population parameters than will smaller samples. Yet, increasing the size of the sample beyond a certain point results in very little improvement and also, additional units bring additional expense. Thus you must balance the need for accuracy against the need to control costs. Further, with very large samples, the quality of the data may even decrease. Note that the relationship between sample size and accuracy applies only to probability samples. Sample size for nonprobability sampling must be governed by other considerations, such as the opportunity to get an in-depth understanding of a problem and possible solutions.

As you think about sample size keep in mind the study's purpose. Unless its purpose is well defined and the validity of your measurements has been established, conducting a preliminary study with a small sample may be more efficient than spending resources on a larger sample.

NONSAMPLING ERRORS

Sampling errors come about when we draw a sample rather than study the entire population. If a probability sampling method has been used, the resulting error can be estimated. This is the powerful advantage of probability sampling. But other types of errors can cause a sample statistic to be inaccurate. Nonsampling errors result from flaws in the sampling design or from faulty implementation. To reduce nonsampling errors you must attend carefully to the selection of the sampling frame, the implementation of the design, and the quality of the measures.

Nonsampling errors are serious, and their impact on the results of a study may be unknown. Taking a larger sample may not decrease nonsampling error; the errors may actually increase with a larger sample. For example, coding and transcription errors may increase if staff rushes to complete a large number of forms.

Similarly, a large sample may result in fewer attempts to reach respondents who are not at home, thus increasing the nonresponse rate.

If members of the sample who respond are consistently different from nonrespondents, the sample will be biased. Other nonsampling errors include unreliable or invalid measurements, mistakes in recording and transcribing data, and failure to follow up on nonrespondents. If a sampling frame excludes certain members of a subgroup, such as low- or high-income individuals, substantial bias may be built into the sample.

CONTACTING PEOPLE TO GET INFORMATION

You can collect data using mail surveys, telephone interviews, e-mail surveys, Web-based surveys, in-person interviews, or a combination of these. To decide which to use you may weigh time, cost, and your need for a probability sample. To get a probability sample you need to consider if you can find an appropriate sampling frame, contact sample members, and encourage their responses. Your well-designed probability sample may become a nonprobability sample if you have a poor sampling frame, are unable to contact members of the sample, or have a low response rate.

One of the first things to ask yourself is why will a subject answer your questions. You may find that clients, donors, or staff will respond if they see the value of the study and your request isn't burdensome. The general public may not be so inclined. Survey response rates have declined substantially in recent years, which is attributed both to less success in contacting respondents and to more people refusing to participate. Low response rates, the percentage of the sample that responds, increase costs and call into question accuracy of results. Without careful planning, you risk a low response rate and wasting your time.

The next question you may ask yourself about contacting your subjects is "what will I say?" While the idea of contacting strangers to ask survey or interview questions may seem daunting, it's actually pretty easy once you have a planned study. Much of what you will say when contacting subjects can be taken from your informed consent form discussed in Chapter 3. When you contact subjects, the most important thing to do is to tell them who you are, what you want from them, how long it will take, and what they will get out of it. Most people are very busy but many will take the time to help you if the time seems reasonable and the study worthwhile.

In this chapter we focus on traditional data collection strategies and what you will want to consider in choosing a strategy. We do not cover the specific details of how to conduct a survey: there are excellent "how to" guides that you should consult if you plan a specific study. We spend little time on the opportunities for and challenges to data collection, such as identifying and contacting potential respondents, offered by new communications technologies. Research on new technologies is in its infancy, too soon for us to know how they can affect your work.[3]

---

[3]A place to start exploring new technologies is *Envisioning the Survey Interview of the Future*, F. G. Conrad and Michael F. Schober, eds. (Hoboken: Wiley-Interscience, 2008).

For instance, when considering a telephone survey you may want to investigate including cell phones in your sample. In addition, with the availability of inexpensive online programs to design and administer online surveys and the diffusion of Internet access, you may consider conducting an online survey, which opens up opportunities to include graphics and even video clips. As the costs and logistics of in-person interviewing climb, you may explore opportunities to conduct interviews and focus groups via video conferencing. To take advantage of new technologies and their impact may require you to keep abreast of research being conducted by survey organizations, such as the Institute for Social Research at the University of Michigan.

Mail Surveys

Mail surveys come to mind when we think about self-administered surveys. As is true of all surveying techniques, mail surveys have their unique advantages and disadvantages. All surveys require time, as you write questions, design a questionnaire, and find a sampling frame with contact information. Mail surveys consume additional time for mailing, delivering, and returning surveys. Of course, things may not go as anticipated. The survey may remain in a mailbox unattended or sit on someone's desk. You may not know why a survey was not returned. Has the subject moved? Was he out of town for an extended period? Did she feel that the questions did not apply to her?

On the other hand, mail surveys may be less expensive than telephone and in-person surveys. You do not have to worry about finding respondents at home or in their office or contacting them at an inconvenient time. Respondents can answer the questionnaire when they want and locate any needed information. Mail surveys may work especially well if you are asking highly motivated people to answer questions about a subject that they care deeply about. An executive director may be happy to answer questions about partnerships with government—partnerships that may provide income and community status, but at the same time increase scrutiny of the agency. Employees may answer questions about benefits either to vent their feelings or in the hope that their answers will influence future management decisions.

Not everyone is willing to spend time answering a survey. Mail surveys normally have low response rates. No one is there to urge the respondent to complete it or to explain an unclear item. A respondent may put the questionnaire aside and never get around to answering it. Easily answered mail surveys should get a higher response rate. Consequently, to make mail surveys easier to complete, consider relying on closed-ended questions and ask the respondent to check a response from a list of options.

The following factors, which have been shown to affect the response rate of mail surveys, can be applied to other types of surveys.

A. Sampling Frame
 1. *Accuracy.* An out-of-date sampling frame will reduce responses.
 2. *Relevance.* Recipients may not respond if they believe they do not belong to the target population.

B. Questionnaire Design.
  1. *Length.* Shorter surveys usually have higher response rates.
  2. *Layout and Format.* The questionnaire should be easy to read and easy to fill out, with items sequenced logically. It should look attractive and of high quality.
C. Delivery and Return
  1. *Prenotice.* Sending a message in advance to respondents telling them that they will receive a questionnaire and something about it usually improves response rate.
  2. *Cover Letter.* Including a cover letter with the questionnaire to explain the reasons for the study, its importance, the importance of the respondent's participation, and an offer to share results typically increases response rate. Response rates are improved if the letter includes an endorsement from someone known to and respected by the respondent.
  3. *Return Envelope.* Including a stamped and addressed envelope markedly improves response rates.
  4. *Follow-up.* Following up with a reminder two or more times improves response rates. Including another copy of the questionnaire with the follow-up improves rates even more. (Note: too many follow-ups may compromise voluntary participation.)
  5. *Incentives.* Including incentives such as discount coupons, donations to a charity, participation in a lottery, or a token such as a pen can improve response rates.

A pilot test (or dress rehearsal) of the survey and its administration should identify how well the above advice works for you.

Internet Surveys

Internet surveys may be easily and inexpensively implemented. Similar to mail surveys Internet surveys should be easy to answer and return. You can use software such as SurveyMonkey, Survey-Gizmo, or Zoomerang to format a questionnaire, post it on a Web site, gather responses, and report the data on a spreadsheet. Although the software makes your work easier, the quality of the survey depends on the quality of the questions and your sample. The surveys are typically posted on the Web or sent as part of an e-mail message, so you don't have to budget for postage or printing. You may have the data go directly into a database for later analysis. This allows you to avoid entering data from paper forms, so you don't have to budget for data entry or verifying the accuracy of the entered data.

An e-mail message may be sent to sample members. The e-mail message should include the same information included in a mail survey cover letter and invite the recipients to complete the survey. They may be given the Web address of the online survey. Alternatively, the message may include a direct link to the survey. To control access, which is recommended to prevent multiple responses, a password or access code may be provided.

Less commonly, a survey is sent as an e-mail attachment. The respondent opens the survey, completes it, and returns it either as an attachment or through regular mail. Such e-mail surveys, however, may raise greater concerns about confidentiality, security of responses, and viruses than online Web-based surveys.

Sampling problems are particularly challenging for Internet surveys. Not everyone has equal access to the Internet and we cannot assume that a survey using the Internet will reach a representative sample. The first challenge is to find a sampling frame. Directories with e-mail addresses are available for some populations, such as members of professional associations and organization staff. This is less true of clients, customers, and the general public. If a sampling frame with e-mail addresses is available, you can choose a sample and e-mail the sample members. E-mail addresses can go quickly out of date. Sometimes a misdirected message will be returned; other times it may get lost in cyberspace. Again, respondents may delete messages from unknown sources or direct them to a spam file.

If a list of e-mail addresses for respondents does not exist, accessing the target population becomes more difficult. You may publicize the survey through Listservs, newsletters, newsgroups, and social networks. However, it is difficult to demonstrate that a true probability sample was select if a survey is disseminated through many channels. Further, you may be unable to identify the represented population. Since you do not know how many people other than those in the target population replied, you cannot calculate the response rate. On the other hand this strategy may approximate a snowball sample and help recruit difficult-to-identify populations.

Telephone Surveys

If you have a sampling frame with telephone numbers you may prefer a telephone survey. It has a faster turnaround time than a mail survey and you can track the reasons for nonresponses, such as inability to contact, not part of the sample population, or a refusal. Telephone interviewing has largely replaced in-person, structured interviews as the latter has become less feasible. Interview time and travel is expensive. Both residents and interviewers may have safety concerns. Diverse lifestyles mean that people are more difficult to contact. Telephone interviewing has the added benefit of providing a rapid response from conceiving an idea through to the reporting of results. Telephone surveys allow interviewers to cover a wide geographic area, and several interviewers can work from the same location.

You can conduct the survey yourself, or you may hire a vendor. When conducting the survey yourself, remember to pretest it and call potential respondents to make sure they understand the questions and stay engaged throughout the interview. Well-trained interviewers may be effective in connecting with potential respondents, getting them to participate, and moving through the survey efficiently. They may use interviewing software with the interview questions and responses appearing on a computer screen. The interviewer reads the questions and enters the responses as the respondent answers the questions. The answers go directly into a database for analysis.

Telephone surveys have been especially valuable in collecting data from the general public. Random digit dialing (RDD) has been the preferred method for collecting data from homes with landlines. A sample is created by linking a telephone exchange with a random number; for example, a local 313 exchange could be completed by adding a random number between 0000 and 9999. RRD does not require a sampling frame and it can reach homes with unlisted numbers. However, other problems remain. Households with caller-id may not even answer. People who live in households with more than one number, which are typically more affluent than the general population, have a higher probability of being selected.

As cell phone–only households become more common, survey organizations are investigating using RDD to survey cell phone samples. One study found that cell phone interviews cost over twice as much as landline-based interviews, have higher rates of refusals, and many more ineligible contacts.[4] Interestingly enough, researchers are starting to explore address-based sampling (ABS) as a complement or as an alternative to RDD. An investigator draws a sample from a list of residential housing unit addresses and the sample receives the survey in the mail. Preliminary research has found that RDD cost 12 percent more than ABS, which included a postcard reminder and resending the questionnaire. Although ABS and RDD cannot substitute directly for one another (e.g., ABS questionnaires may have to be shorter), both approaches may be combined to collect more representative data.[5]

In-Person Interviewing

In-person, face-to-face interviews enable researchers to obtain detailed information and ask complicated, in-depth, and sensitive questions. A few major national surveys, most notably the Current Population Studies conducted by the U.S. Census Bureau, conduct in-person surveys with a probability sample of American adults. In-person interviewing is a basic tool of qualitative researchers, who may conduct semi-structured interviews and focus groups. To arrange interviews with specific individuals a variety of strategies can be used: asking an intermediary to introduce you to potential respondents or introducing yourself through a letter, e-mail message, or telephone call. To ensure a high response rate and valid data, potential respondents are typically contacted ahead of time to schedule a time and place for the interview.

Unlike telephone interviewing, where interviewers are closely supervised and monitored, in-person interviewers are on their own. They must be well trained. They must be able to explain the purpose of the study and the meaning of the questions. They have to be able to follow the interview protocol, including knowing how to handle sensitive or confidential information. Some in-person interviews involve electronic equipment, such as video and audio recorders and computers, in which case interviewers must be able to operate the equipment and know what to do in the case of breakdowns.

---

[4]The Pew Research Center for People and the Press, "News Release: 'National polls not undermined by Growing Cell-Only Populations' " posted at http://people-press.org/reports/pdf/276.pdf (2006). Accessed January 17, 2010.

[5]M. W. Link et al., A comparison of address-based sampling (ABS) versus Random Digit Dialing (RDD) for General Population Surveys, *Public Opinion Quarterly* (2008), 72(1): 6–27.

CONCLUDING OBSERVATIONS

The topics of sample design and data collection techniques are much more interrelated than you may have initially realized. Theoretically, designing a probability sample seems straightforward. The first challenge is identifying a sampling frame. Oftentimes sampling frames do not exist or cannot be used for the proposed survey technique. While random digit dialing seemed to resolve the problem of nonexistent sampling frames, the spread of cell phones and consumer resistance to unsolicited calls have made telephone surveying less attractive. Furthermore, telephones may not be the best medium for long surveys or ones with many open-ended questions.

A closely related problem is the response rate. A poor response rate undermines the strongest sample design. Mail surveys have generally had the lowest response rate. In-person surveys should have a high response rate, but their cost and other logistic concerns limit their use. Internet surveys should potentially get a good response rate if the survey is engaging and can be answered easily, but they are limited in their ability to reach the general population.

A third consideration is the type of questions. Mail surveys normally work best with easily answered, closed-ended questions. Although there are challenges, telephone surveys can work reasonably well with a variety of questions as long as their meaning can be easily understood. Internet surveys seem most suited for closed-ended questions, although they can also handle open-ended questions. In-person surveys are best for complicated questions, ones that require detailed or lengthy information, or ones that ask a respondent to view a video or listen to a recording.

Research is currently being conducted on how to take advantage of new technologies and databases. Some of the research, such as using ABS, may overcome the weaknesses of existing methods. Consider the possibilities of including graphics and videos on Internet surveys or to conduct focus groups using video conferencing. You could explore how residents respond to a new marketing campaign or get diverse viewpoints on how to deliver services more effectively.

Perhaps the most important lesson is to recognize that a well-designed probability sample if poorly implemented can become a nonprobability sample. Instead of a high level of accuracy, the data may have serious biases, eliminating your ability to generalize to the target population.

RECOMMENDED RESOURCES

Arlene Fink, *How to Conduct Surveys: A Step by Step Guide*, Fourth Edition (Thousand Oaks, CA: Sage, 2009).

Fred Fowler, Fred, *Survey Research. Methods*, Sage Applied Social Science Research Methods Series, Fourth Edition (Thousand Oaks, CA: Sage, 2009).

Gary T. Henry, *Practical Sampling* (Newbury Park, CA: Sage Publications, 1990) is a short, readable treatment of sampling with well-developed examples from actual applied research.

Public Opinion Quarterly, the journal of the American Association for Public Opinion Research, is highly recommended to keep up to date with advances in collecting survey information.

CHAPTER 6 EXERCISES

There are four exercises for Chapter 6. The exercises develop your competence in interpreting and applying sampling concepts and in methods to select subjects for surveys.

- Exercise 6.1 Learning from Practice asks you to find a research article and identify and assess its sampling and data collection strategies.

- Exercise 6.2 The Long Street History Museum asks you to design a sampling and data collection strategy to solicit the opinions of museum members about proposed changes as the museum deals with a loss of funds.
- Exercise 6.3 Health Care for Children and Adolescents asks you to design a representative sample from paper medical records stored in filing cabinets.
- Exercise 6.4 Own Your Own asks you to state a research question about an issue at your job or on your internship and design a sample which will help you answer the question.

EXERCISE 6.1   Learning from Practice

Locate an empirical article in a journal or on the Web. Sources for locating studies include http://www.gallup.com/poll/; http://www.ropercenter.uconn.edu; http://www.cadsr.udel.edu/default.cfm; http://www.gao.gov/).

Section A: Getting Started

1. Identify the population, sampling frame, and unit of analysis.
2. Describe the sampling design.
3. How were the data collected (telephone survey, internet survey, or some other way)? What was the response rate?
4. What are the strengths and weaknesses of the design?

Section B: Class Discussion

1. Based on your limited sample of studies (Section A) create a list "What you should say about your sample." Include what information should be included about the
   a. population
   b. sampling design
   c. response rate
   d. sample size

EXERCISE 6.2   The Long Street History Museum

Scenario

The Long Street History Museum relies on several sources to fund its operations. These include board member contributions, donations from the public, entrance fees, membership fees, and grants from both public and nonprofit organizations. Because of a slow economy, the city will no longer provide grants unless the museum merges with another organization in the city. The museum's executive director plans to propose two options to the board of directors: (1) merge with another organization or (2) increase fund raising efforts in order to operate without city grant money. Before doing so, however, she wants to know what the museum members think about the options. She has contacted you to help her find out what their opinions are.

Section A: Getting Started

1. The museum has over 2,000 members. You have been asked to sample 300 members to learn their opinions about the director's proposals.
   a. What questions would you want to ask the director prior to designing the sample?
   b. How would you select the 300 respondents and collect information from them?
   c. Justify your choice of sampling design.
2. Identify your sample's (question 1b)
   a. population
   b. parameter
   c. category of sample (probability or nonpropability)
   d. unit of analysis
3. Identify two different methods of contacting respondents and obtaining information and discuss the advantages and disadvantages of each.
4. Develop a list of questions to ask the respondents.
5. Assume that the executive director also wants to learn what some board of directors members think about her proposals before presenting them to the full board. The board consists of 18 people who are of diverse races, ages, and professions. The board is equally split by gender. Design a sampling and data collection procedure for gathering information from board members.
6. A colleague suggests conducting interviews to explore the feasibility and impact of merging with an organization. Describe whom you would suggest interviewing. What sampling design would you use to identify and contact participants?

EXERCISE 6.3   Health Care for Children and Adolescents

Scenario

A health care organization providing specialized services to children and adolescents wants to destroy thousands of old medical records. It is required by its funding agreements with the state government to save a representative sample of medical records. The records, all in paper files, are stored in 20 metal filing cabinets. Each cabinet holds 1,000 records organized by year. The files for each year are numbered sequentially.

Small Group and Class Exercise

1. Small groups should design a probability sample using each of the following designs: simple random; systematic; stratified; multistage-cluster.
2. Each group should decide which design to recommend and justify its choice.
3. Each group should present its recommended design to the class and answer questions about its design and how the group would implement it.
4. At the end of the exercise each group should prepare a short memo evaluating the strengths and weaknesses of each design and recommend one.

EXERCISE 6.4 On Your Own

1. Identify an issue related to your place of work or internship and pose one or more research questions regarding the issue.
2. Develop a sampling design for a study that would obtain information to answer the research questions. In developing your responses to the following items, use plain English to explain your methods to your supervisor and other stakeholders.
   a. Identify the study population, sampling frame, sampling design, and sample members.
   b. How will surveying or interviewing this sample help answer your research questions?
   c. Justify your use of a nonprobability or probability sample.
   d. How will the availability of resources affect your choice of design?

Questions and Questionnaire for Surveys and Interviews

Questionnaires are a basic tool of an effective manager. The questions give life to measures that describe program performance, service utilization, or citizen satisfaction. Program personnel and policy makers use the answers to plan, evaluate, and monitor programs. By this point in your life you have probably answered several surveys and filled out numerous forms. If you answered quickly you may have never considered how much work went into making the survey user friendly and the form easy to answer. Writing clear questions, whether you will put them on a form or ask them in a focus group, isn't easy. You must take care in their wording. People can only answer the question that they are asked, not the one you meant to ask.

Respondents must believe that answering the questions is worthwhile. For example, you may have received requests for information that is not readily available or for opinions about policies or programs that don't interest you. You may resent the intrusion on your time and the assumption that you are willing to sacrifice energy or effort to answer something that is of no apparent value to you. Similarly, when respondents are asked to report data that they do not value, they may estimate or make up answers to get a form off their desk. Alternatively, they may stop answering if they find that a survey is too time consuming, confusing, or seems unlikely to produce useful information.

In this chapter we focus separately on two different modes of designing questions. One mode designs questions for forms and surveys The questions tend to be closed-ended, are meant to elicit specific answers, and involve little or no interaction between the respondent and the investigator. The other mode, building on survey techniques, designs questions for semi-structured interviews and focus groups. These questions are open-ended, are meant to elicit detailed answers, and involve interaction between the respondent and the interviewer.

QUESTIONS FOR SURVEYS AND FORMS

To decide what questions to ask you need to learn what questions program stakeholders want answered and how they plan to use the answers. You would not want to exclude questions needed to make a decision, nor would you want to ask unnecessary questions. Unneeded questions increase costs and burden respondents. Questions may be written to gather information on

- facts
- behavior
- knowledge
- opinions
- attitudes
- motives

Factual questions ask respondents for demographic or other objective information. For example, organizations may be asked the number of full-time employees, part-time employees, and volunteers. *Behavior questions* ask respondents about things they do or have done. Respondents may be asked if they have contributed to a charity and if they have, how much they contributed. Behavior questions do not have to yield a number. For example, an organization may be asked "to describe how you manage your volunteers."

Knowledge questions determine what a person knows about a topic and the accuracy of his or her knowledge. Think about questions such as "Which has more calories: a ¼ pound hamburger or 8 ounces of flavored yogurt?" or "Do nonprofits have to pay taxes?" Calorie counts and tax policies are facts, but you may ask these questions to identify what respondents know. You may ask knowledge questions at the end of a training program to see what participants learned. If you plan to conduct a public opinion poll you might use knowledge questions to identify topics about which the public is ignorant or misinformed.

Opinion questions ask people what they think about an issue or event. *Attitude questions* focus on more stable, underlying beliefs or ways of looking at things, such as attitudes toward abortion, the death penalty, or race relations. In practice, however, the terms opinion and attitude are used interchangeably. So, you need not debate whether you are tapping an opinion or attitude. *Motive questions* ask respondents why they behave in a certain manner or hold certain opinions or attitudes. To learn why people eat what they do, you might ask respondents to rate the level of their agreement or disagreement with statements such as[1]

- It is important that the food I eat on a typical day is easy to prepare.
- It is important that the food I eat on a typical day is low in calories.
- It is important that the food I eat on a typical day makes me feel good.

[1]These statements are from an extensive measure of what motivates food choices and can be found in A. Steptoe, T. M. Pollard, and Wardle, "Development of a Measure of the Motives Underlying the Selection of Food: The Food Choice Questionnaire," *Appetite* (1995), 25:267–284.

Knowing what information a question produces avoids collecting the "wrong" data. Three common errors are: assuming that perceptions reflect reality, that intentions reflect what a person will actually do, and that satisfaction measures quality. While knowing how a person judges the quality of his diet (a perception) may be useful, it doesn't tell us if his diet is actually healthy or unhealthy. Knowing how many people say they would eat more salads when they eat fast food if salads were available (an intention) may not correspond with the number of people who actually order salads. Asking the audience at a presentation on healthy eating to rate the presentation (satisfaction) does not indicate if they learned anything, if they used what they learned, or how nutritionists would judge the presentation. Measures of people's perceptions, intentions, and satisfaction may not be operationally valid if the measures were intended to indicate the actual state of affairs, future behavior, or program quality.

Each question should be worded so that a respondent understands what it means and answers honestly. A question should be clear, specific, and short. Clear questions contain words in common usage and are not subject to different interpretations. Imagine a question asking, "Do you own stock?" People can interpret stock as an investment instrument, livestock, or broth. Questions should avoid using abbreviations that respondents may be unfamiliar with or easily misunderstand.

You must know exactly what you want to measure and communicate it clearly. Consider how respondents may interpret questions asking if they work and how much they work. A teenager might define work as including homework, household chores, walking a neighbor's dog, or occasional babysitting. Women, in particular, may feel that housework should be included as "work." Retirees who regularly volunteer may believe their volunteer hours constitute work. To avoid confusion you may define key terms as part of your instructions. For example, consider a survey that asks employees about an organization's management. The following list may have been included in the instructions:

- *Senior executives* means the president/CEO/executive director, associate and assistant directors, and other senior administrators.
- *Supervisor* means the person to whom you report on a daily basis.
- *Work group* means the coworkers in your work unit.

Alternatively you may incorporate the definition of a term into the body of the question. For example, the following question clarifies to whom "your supervisor" refers:

How often, if at all, does your supervisor, that is, the person to whom you report on a daily basis, require that changes be made to the findings or conclusions of a report?

Questions should be specific about the time, place, or amount involved. For instance, to estimate clinic usage, investigators may ask, "How many visits did you make to the health clinic in Middletown between September 1 and December 31 of this year?" If "Middletown" did not appear in the question, respondents might count out-of-town clinic visits. If the question said "recently" instead of "between September 1 and December 31 of this year," each respondent could answer the question with a different time interval in mind. Questions asking respondents how often they do something should cover a limited time span. Some respondents may think of the number of visits within the last month; others may think of the number of visits within the past year. Similarly, someone may accurately remember how many times he used a library in the past month. If he is asked about the past year, he may multiply the number of visits in the past month by 12. He is unlikely to adjust his estimate if during the month he used the library more than usual or less than usual.

Short questions are normally clear. When listening to a long question or reading it, a respondent may pay attention to only a part of the question. A large number of words or dependent clauses may distract a respondent and she may not take the time to figure out exactly what is being asked.

Closed-ended Questions

Survey questions are characterized as being closed-ended or open-ended. Closed-ended questions include a list of responses; open-ended questions do not. The list of responses in closed-ended questions should include all probable responses, and the responses should be mutually exclusive. A common error is to ask a respondent to indicate if she is "single," "living with a partner," "married," "separated," "divorced," or "widowed." In such cases, it is not uncommon for a person to correctly check several categories. The common solution is to ask for "Current marital status" and specify single as "Single (never married)." If relevant responses are not included the item's reliability is compromised. A survey sent to graduates of a training program might ask, "What is your main job function?" The list may include a variety of activities, but fail to include "Not working" as a possibility. A respondent faced with a lack of appropriate choices may skip the question, check a less than accurate response, or find someway to record an alternate response. On the other hand an exhaustive list may be infeasible or burdensome. Listing "Other" as a choice may lessen the problems caused by a respondent not seeing an appropriate answer.

In addition to "Other," response choices such as "No opinion," "Not sure," or "Not applicable" lessen the probability that a respondent will give an inaccurate or random response. In the case of opinion questions, some investigators do not want to give neutral choices; they assume that everyone has an opinion. Questions that do not allow respondents to give a neutral response or no opinion are called *forced choice questions*. Of course, you cannot force respondents to choose. Some may pen in their own answers or ignore the question. Forced-choice questions may be particularly annoying to recipients of online surveys, who may not be able to proceed until they answer all questions on the screen. Faced with irrelevant responses or stuck on a computer screen a respondent may check any old answer or abandon the survey entirely.

The list of responses may indicate how precise an answer is expected. Instead of using indefinite terms such as Always, Often, Sometimes, and Never, you should consider giving numerical ranges. For example, to ask volunteers how often they used skills they learned in training, the responses might be "Every time I volunteer," "At least half the time I volunteer," "Less than half the time I volunteer," and "Never." A response list may serve as a frame of reference, and a respondent may alter his answers if they seem atypical. If you wish to avoid suggesting appropriate answers you may ask respondents to fill in their answer.

You may ask respondents to choose just one response or you may ask them to check All that apply. For example, you might ask, "Indicate which communication technology you use most often" or "What communication technologies do you use (check all that apply)?" In practice respondents may not check "All that apply." An unchecked item may (1) not apply or (2) have been overlooked. A better strategy is to ask respondents to check separately an appropriate response for each listed item. A format frequently found on online surveys is to present a matrix with the questions, items, or statements listed on the left-hand side and response choices along the top. The following illustrates a matrix format:

How Much Time Do You Use Each Communications Tool?	Often	Throughout the Day	At Least Once a Day	At Least Once a Week	Less than Weekly	I Do Not Use This Technology
Internet						
Mobile Telephone						
Personal Digital Assistant (PDA)						

You may ask respondents to rank or rate items. Because we have encountered response problems when we have asked for rankings, we do not recommend this practice. Assume you ask respondents to rank how important they consider seven listed communications technologies (including social networks) and give a 1 to the most important and a 7 to the least important. Inevitably some respondents will rank each technology with a "1." A few respondents may appear to reverse the scale, giving a "7" to the most important technology and a "1" to the least important. Some will rank a few technologies and ignore the others. A similar format asks people to assign percentages to how much time each week they used each technology. A surprising number of people make the mistake of assigning values that add up to well over 100 percent.

Questions for Self-Reports of Behavior and Opinions. To learn about people's attitudes, opinions, motives, and behavior requires investigators to question individuals directly to elicit self-reports. However, the accuracy of such self-reports should be viewed with skepticism. People tend to give socially desirable answers, memories are fallible, and intentions may be transient. Socially desirable answers reflect how respondents think they should act or what they should believe, and not how they actually act or what they actually believe. As an interviewer and respondent interact this tendency on the part of the respondent may increase.

The accuracy of the answers also depends on how respondents store and retrieve information. Consider self-reports of voting. From time to time,

---

[2]Most of the information on validating voting data is from S. Presser and M. Traaugott, "Correlated Response Errors," *Public Opinion Quarterly* (Spring 1992), 56:77–86. For a discussion on voter records as a source of error, see P. R. Abramson and W. Claggett, "The Quality of Record Keeping and Racial Differences in Validated Turnout," *Journal of Politics* (August 1992), 54:871–880.

researchers check self-reports against voting records.[2] They have found that roughly 15 percent of the people surveyed after an election misreport whether or not they voted. Three reasons have been proposed to explain why the misreporting occurs. First, respondents who usually vote may have forgotten that they did not vote in the election being studied. Second, voting records may be in error. For example, a check of records may miss a voter whose name on the voters' rolls differs from the name she gave to the surveyor. Third, misreporters may wish to give a socially desirable answer. The direction of misreports is skewed. A higher percentage of respondents report having voted for the winning candidate than was indicated by actual votes.

Uninformed or disinterested respondents may express an opinion about something they have no knowledge of or interest in. To discourage uninformed responses, Don't know may be included as a response choice. A "Don't know" response may identify a respondent who does not have an opinion, a respondent who does not want to state his position, or a respondent who has thought about the issue but has yet to reach a firm opinion. During telephone or in-person interviews some respondents may become distracted while hearing a question or have trouble understanding it. They may say they "Don't know" to avoid the trouble of having the question repeated or trying to decipher its meaning. Investigators must examine any question that generates a high proportion of "Don't knows" to make sure it is not ambiguous or otherwise flawed.

To separate disinterested respondents from other respondents, you may first ask a question with a dichotomous response and then a question with responses that measure intensity. For example, if you wanted to measure popular support for collecting property taxes from churches and other houses of worship you might ask

1. Do you support or oppose having houses of worship, such as churches, pay property taxes?
   - ___ Support
   - ___ Oppose
   - ___ Don't know
2. How interested are you on the issue of having houses of worship pay property taxes?
   - ___ Very interested
   - ___ Somewhat interested
   - ___ Somewhat disinterested
   - ___ Not at all interested

Try to avoid having all statements go in the same direction. For example, for some statements an "Agree" would show satisfaction with a program and a "Disagree" would show dissatisfaction. The danger is that a respondent may answer each question with the same response regardless of content. You should avoid wording that can be easily misread or misunderstood. For example a respondent may overlook a key word such as *not*. Again, respondents may become annoyed or confused if questions seem to change direction capriciously. The questions that follow ask clients to rate agency staff. The questions illustrate how questions can be worded to change direction effectively. Respondents answer each statement on a scale ranging from Strongly agree to Strongly disagree.

1. The agency staff can answer my questions.
2. The agency staff is polite when I telephone.
3. The agency staff is accessible by telephone.
4. The agency staff is rude when questions are asked.
5. The agency staff interrupts me when I talk with them.

A person who is satisfied with agency staff should agree to items 1 through 3 and disagree with items 4 and 5. The shift in direction is achieved without resorting to awkward wording. As shown in these questions, a respondent should be able to quickly and accurately grasp the meaning of each question.

Surveys that ask personal questions may include questions that some respondents consider private or offensive. Each of us has feelings about who we are, what we believe, and how we behave. Thus, questions that seem likely to create discomfort should be asked only after careful consideration. Otherwise, respondents may give untruthful answers or if they are sufficiently uncomfortable they may not complete the survey. Some topics and the related questions may raise ethical concerns that the answers may harm the subjects. Some questions themselves may evoke intense feelings, such as questions about experience with domestic violence. Some questions, such as those about drug use, may ask about illegal behavior.

You should not assume that your perceptions of what questions will cause discomfort is correct. For example, an acquaintance from India observed that Americans seem to talk freely about their feelings toward family members but refuse to discuss their salaries. Conversely, he noted that Indians talk about their earnings openly but virtually never discuss their feelings about family members.

Stating a question so that it suggests to the respondent that any possible answer is acceptable may reduce discomfort. For example, a question may start with the phrase, "Some people find . . ." Question order also can reduce the discomfort associated with a question. Potentially threatening questions should not be asked at the beginning of a questionnaire, where they may make a respondent suspicious and less willing to cooperate. Nor should they be placed at the end, where they may cause a person to finish a survey feeling anxious and wondering if the survey had a hidden purpose.

Biased Questions: A *biased question* encourages respondents to give one answer over another. One source of bias is to include a positive or negative prompt in the question. For example, the question "Do you agree that charities should pay property taxes" may seem to encourage a respondent to answer "agree." You might restate the question as "Do you agree or disagree that charities should pay property taxes?"

A common source of bias is to ask two questions in one. These questions are easily identified by the conjunction *and*. Consider the question, "Are program staff friendly and knowledgeable about services?" A person may find the staff friendly, but not knowledgeable. Such a respondent may check "Agree," possibly to be agreeable, but her answer wouldn't be accurate. Conversely, it is not at all clear how an investigator should interpret a "Disagree." Did the respondent find the staff friendly but not knowledgeable or did she consider the staff both unfriendly and not knowledgeable?

Questions may include an assumption that influences responses. For example, consider a question asking respondents whether they agree or disagree that "teacher salaries should be raised to improve student achievement." The question implies that raising teacher salaries will increase student achievement. Respondents concerned about student achievement may be led to voice support for raising teaching salaries solely because the question linked salaries with student achievement.

Loaded questions have words or phrases that evoke a strong positive or negative response. Such words or phrases lead a person to ignore the major content of the question. For example, labeling a proposed policy as "liberal," "permissive," or "bureaucratic" will tend to measure respondents' reactions to these words and not their general opinions about the subject at hand.

Rating scales can also be a source of bias. Recall from our discussion of reliability in Chapter 2 that excluding a relevant response from a list may create a bias. A rating scale with a disproportionate number of positive (or negative) ratings is also biased. For example, a list that asks a respondent to rate a service as "Excellent," "Good," "Satisfactory," or "Poor" has a positive bias. Three of the four responses are positive.

Other Considerations: The best questions ask about the here and now, rather than about past behaviors or future intentions. People may not remember past behaviors; recall our comments about misreporting voting behavior. Similarly, people may not be aware of factors that will affect their future decisions. Residents who say that they would take public transportation may decide not to once it is available if it is too expensive or inconvenient, or a requested training course may prove to be less attractive once potential trainees read the description. Questions that ask respondents how much they would be willing to pay for something are particularly sensitive to wording. If the respondent feels that her answer will affect service cost, she may cite an amount lower than what she would pay. Consider a question that asks people whether they would pay $15 for a tennis license to play on the city's public courts. A tennis player, who may be quite willing to pay $15, might try to influence a policy decision by answering "no."

Occasionally a response list may be entirely wrong for a question. This is apt to happen if you assemble a questionnaire by cutting and pasting response lists You may inadvertently insert the wrong list. For example, you may paste the response list Strongly agree, Agree, Neither agree nor disagree, Disagree, or Strongly disagree to the question "How knowledgeable is program staff?" None of the choices, of course, answers the question.

Open-ended Questions

Open-ended questions require the respondent to answer in her own words. Open-ended questions are included on surveys for five reasons. (1) They help you identify the range of possible responses. (2) They avoid biases that a list of responses can introduce. (3) They yield rich, detailed comments. (4) They give respondents a chance to elaborate on an answer to a closed-ended question. (5) They may be easier to answer than scanning down a long list of possible responses. For example, the question, "In what state do you live?" is easier to answer by writing a state name than by looking for the state on a list.

Open-ended questions are particularly helpful as a survey is being designed. The answers suggest appropriate response categories for the survey. In political polling the practice of progressing from open-ended to closed-ended questions is well established. Long before American political parties and voters express any formal preference for a presidential candidate, pollsters ask respondents to name whom they prefer as the next president. Pollsters can identify candidates with early support while avoiding the possibility that respondents simply choose a familiar name, in which case the poll would be measuring name recognition rather than support. Long before Election Day open-ended questions disappear and are replaced by closed-ended questions that ask about actual, active candidates. A similar pattern occurs if you ask stakeholders about suggested policy and program alternatives.

Answers to open-ended questions provide the rich detail that puts a mass of collected data into context. If a policy is being planned or a program evaluated, a respondent's comment may improve an investigator's understanding about how the public views the policy or how consumers experience a program. You may incorporate the comments into a report, which adds to its readability and keeps the interest of less quantitative readers.

Open-ended questions have three major drawbacks. First, open-ended questions require motivated respondents. Second, they complicate data compilation. Third, they increase costs, specifically the cost of training staff to collect and compile the information and the cost of verifying the reliability of the summarized information.

Respondents with little time, minimal interest in the topic, or limited communication skills may ignore open-ended questions. We assume that investigators usually have more at stake in a survey and its findings than the individual respondents. Thus, investigators overestimate respondents' willingness to provide detailed answers. We assume that, generally respondents avoid questions that require more than just a few words to answer. This assumption may not apply if the respondent knows that the survey findings will affect a decision important to him.

Categorizing and counting the answers to open-ended questions is difficult and time consuming. If you have surveyed relatively few people, you may carefully read each answer and decide how to incorporate it in your analysis. If you have surveyed many people and don't have time to review all the open-ended comments, one strategy you could use is to select a random sample of responses for detailed analysis. Whether you examine a sample or all completed surveys, you will find the task of categorizing the responses accurately and consistently to be challenging. To ensure reliable data, you may rescore a random sample of cases. If you give virtually the same values both times you may assume that the data are reliable. Similarly, if several analysts are categorizing responses, you should check for inter-rater reliability. Have each analyst categorize the same sample of cases and the compare categories the analysts assigned for comparability.

Reviewing the Questionnaire

Beyond the development of questions, other decisions must be made to ensure that your survey is user friendly, encourages accurate and complete responses, and yields useful information. Describing recommended strategies in depth is beyond the scope of this text, but we have created a checklist to guide you in constructing a survey.

Self-Administered Surveys

- The purpose is clearly stated in the introduction.
- Directions on how to answer are clear.
- Recipients who do not belong to the target population are identified, often with a preliminary question. If they are not part of the target population they are told what to do with the survey. For example, they may be asked to return the unanswered survey or to give it to a member of the target population.
- The deadline date and return address appear on the survey.

Interviewer-Administered Survey

- The purpose is clearly stated in the introduction.
- Instructions tell the interviewer how to determine if a potential respondent belongs to the target population and if not how to look for a replacement respondent.
- Directions indicate how to ask questions and record answers.

On all Surveys and Data-Collecting Instruments

- Estimate of how long it will take to answer the questions.
- Define critical terms.
- Do not use abbreviations.
- Do not use conjunctions, such as "and."
- Have adequate and appropriate response choices.
- Request easily accessible data.
- Group questions logically.

Questions on Opinion and Attitude

- Use neutral question wording, such as "Do you favor or oppose . . .?"
- Use balanced responses, for example, an equal number of positive and negative responses.

DESIGNING THE QUESTIONNAIRE

Introductions and First Questions

The introduction and first questions should engage a respondent and encourage his participation. The introduction should be short and to the point. In addition to stating the purpose of the study, the introduction should

- identify the survey's sponsor
- indicate what will be done with the survey information
- indicate whether the responses will be confidential or anonymous

- describe how promises of confidentiality or anonymity will be maintained
- indicate how long the survey should take
- indicate the value of the potential respondent's participation

Practices vary as to the need to specifically remind potential respondents that their participation is voluntary and that they may discontinue answering questions at any time. In the case of mail or telephone surveys, voluntary participation is often implied because subjects usually feel no compunction about ignoring the survey or hanging up the phone. If participation involves possible risks the introduction should identify them. The most likely risk—other than wasting someone's time—is raising painful memories.

The following is an example of an introduction to a self-administered survey.

SAMPLE INTRODUCTION

The Oaks Chamber of Commerce wants to learn how you and others feel about the Town of Oaks as a place to work and live. If the data are to be useful, it is important that you answer each question frankly and honestly. There are no right or wrong answers to these questions.

Your answers to these questions are completely confidential. This survey is being conducted by the state university where they will be analyzed. No one other than the state university researchers will ever have access to your individual answers. So that we can follow up with nonrespondents, a number has been put on this questionnaire that can be matched with your name on a list at the university. It would be appreciated if you would leave this number intact.

Thank you in advance for your cooperation and assistance.

The introduction or the first question should ask if the respondent is a member of the target population. If not, recipients of mail surveys may be asked to return the uncompleted questionnaire. For a telephone survey, you may instruct interviewers to terminate the interview or to see if an eligible subject is available.

The best first questions are those that are easily answered and clearly related to the survey's purpose. With face-to-face interviews, these questions build rapport between the interviewer and the respondent. In a self-administered questionnaire, the questions draw the respondent into making a "psychological commitment" to complete the questionnaire. The first questions keep a potential respondent from putting the survey aside and forgetting it; however, if the survey becomes unduly complex, confusing, or time consuming the benefits of good first questions may be lost. In a telephone survey, the first questions may be important in getting cooperation from a wary respondent who may suspect that the caller is really going to sell him something. Here are examples of opening questions.

Customer Satisfaction Survey of a Supported Employment Program:
Please rate your overall satisfaction with [name] program.
I feel like [name] program has helped me.
I would recommend [name] program to a friend.

> Client Utilization Survey for a Childhood Immunization Clinic:
>
> How did you first hear about this immunization clinic?
>
> Is this the first immunization clinic you have attended?
>
> How did you get to this clinic today?

Filter Questions

Among the questions may be *filter* questions, which separate respondents and direct them to the appropriate parts of a questionnaire. A filter question on a housing survey might be

> Do you own or rent the house in which you now live?
>
> Own (answer question 10)
>
> Rent (answer question 11)

Can you imagine one or more questions that apply to homeowners but not renters? For example, questions asking about mortgages, interest rates, and property taxes would apply to owners but not directly to renters.

Demographic Questions

If you plan to ask demographic questions place them at the end of the survey. By the end the respondent may be less reluctant to answer personal questions. Some investigators get into the habit of asking a standard set of demographic questions, such as questions about age, sex, race, education, and income. These questions may go unanalyzed and contribute little to the study. Surveying consumes resources, including respondent goodwill. Thus, you are well advised to stick to information appropriate to the study and to eliminate questions that have no role in the planned analysis. In addition, too many personal questions may disturb respondents who worry about the anonymity of their answers.

QUESTIONS FOR INTERVIEWS AND FOCUS GROUPS

You conduct surveys to obtain quantifiable information. You may conduct focus groups and hold interviews to get in-depth information and insights about a topic. We refer to these techniques as *qualitative interviewing*. Qualitative interviewing requires that you connect with your subjects, listen well, and be flexible. If you don't have these skills you may find it more practical to conduct a survey.

Qualitative interviewing relies on the rapport and the give and take between you and your subjects. As part of building rapport and getting valuable information you must listen carefully to each answer. The answer to one question will suggest additional questions. Closely related is your ability to maintain flexibility. If your planned questions aren't working you may have to switch gears. The beauty of interviewing is its fluidity and ability to respond to what a subject is saying.

You may erroneously assume that qualitative interviews are just chats and that you can start without having questions in mind. Prior to conducting qualitative interviews, you need to have a carefully designed set of questions. The questions should not be answerable with "yes" or "no" responses. For instance if you want to know about how an economic downturn has impacted the programs offered by a nonprofit organization, you would not simply ask "Has the recent economic downturn hindered your programs?" A respondent's "yes" or "no" omits the detail that is the crux of qualitative interviewing. Consider asking "How has the recent economic downturn affected your programs?" Can you imagine why this question will yield more useful information? Also, note that we substituted "affected" for "hindered," since the latter assumes only a negative impact and the interviewer would miss the opportunity to learn of beneficial impacts.

What follows are guidelines that you will want to keep in mind as you write questions for qualitative interviews.

Is This Question Necessary? Try to stay on task when conducting an interview. Be sure to remember your research question as you write questions and potential probes. Remember, participants have agreed to spend their valuable time with you. Asking questions for the sake of asking questions wastes their time. Therefore, ask yourself: Is this question relevant? Will it help answer the research question? Will it add to my knowledge of the issue at hand? Will it explain any theories or hypothesis I am exploring? Is it simply satisfying my curiosity? Keep in mind that each question adds time to the interview, the transcription, and the data analysis. Each task is time consuming.

Is This Question on Topic? Each question should be related to the research question. While interviewing, you may get off topic—especially if you have a good rapport with your respondent. We are all social creatures and most of us love to talk about ourselves. Therefore, remind yourself of the importance of staying on topic! Remember why you are there and what you are doing.

How Will This Question Be Heard? Remember your subjects. Individuals have their own respective lenses through which they view the world. Commonly, when people are asked why they perform a certain act, they may hear it as a challenge to their decision. A question like "Why did you stop being a vegetarian?" may be heard as an attack on the decision. Therefore, imagine how others will interpret your questions. Just as we urge you to pilot survey questions, be sure to pilot interview questions. Are they clear? Do they get the responses you seek? Does a question prompt the respondent to go into a completely different direction or become defensive? Returning to the example regarding vegetarians, when questioning former vegetarians, a less confrontational question might be, "Why did you decide to include meat in your diet?"

Is This Question Leading? Social desirability is an even greater concern with qualitative research than with quantitative research. You are sitting right there looking at the individual as she answers. That's pressure! The more the questions and the way they are asked can be kept neutral, the better. If you imply how you would like respondents to answer, it is more than likely they will answer that way.

Other Considerations: As previously discussed, open-ended questions generate valuable information. However, even more detail may be necessary to get a clear understanding of the response, so you may want the respondent to elaborate. This is where probes are useful. *Probes* are questions meant to get more detailed answers. Hypotheticals, posing the ideal, and acting as the devil's advocate are three types of probes that you may find useful.

- *Hypotheticals* pose a suppositional circumstance, condition, scenario, or situation. For example, you might ask, "Suppose your organization lost funding for the credit counseling program. What strategies would your organization use to recover?" Such questions encourage the respondent to think beyond what currently is, to what could be.
- *The ideal or the magic wand* asks respondents to think about the perfect scenario. For instance, the mission of many social welfare organizations is to correct a social ill. A probe could ask what is needed to accomplish the mission. You might ask a credit counseling organization, "What has to happen to make sure that all homeowners have the necessary information to avoid foreclosures?" Alternatively, you might ask the respondent what would change if he could "wave a magic wand." You might ask, "If you could wave a magic wand to avoid foreclosures what would you change?" Both strategies provide respondents with the opportunity to think beyond their current situation and explore additional options.
- *Devil's advocate probes* put you in the position of arguing against a cause or position. You should not play the role of a committed opponent, but someone trying to understand the respondent's cause or position. It is important to remain nonconfrontational. Typically, interviewers introduce such probes by saying, "Let me play the devil's advocate." Acting as the devil's advocate allows respondents to consider their responses, justify their current position, fill in missing pieces of information, or think through other options. An example of a devil's advocate probe is "Let me play the devil's advocate and ask why homeowners who knowingly took on too much debt should avoid foreclosure?"

In addition to designing questions you need to think about the interview itself. Two topics worthy of consideration are how you are going to treat silences and handle vague responses.

Your parents may have told you that silence is golden. In the case of interviewing they are right. Often new interviewers are uncomfortable with the silence that follows a question. You may think the respondent does not understand the question. However, often, silence may indicate that a question is particularly poignant or that you have raised a troubling or controversial issue. Silence allows respondents to organize their thoughts and give thoughtful answers. Jumping in with a rephrased question may encourage a superficial answer or one that lacks depth the respondent was pondering. Qualitative interviewers must become comfortable with silence. Silence is your friend in the interview process.

Seeking concreteness is another important technique. Often respondents will answer in general terms. For qualitative research to be meaningful, concreteness and specificity must be achieved. A surefire sign that you are not dealing in specifics is if respondents use words such good, bad, okay, many and do not explain why something is *good* or what is meant by *many*. The terms will not provide the information needed for adequate analysis. When answers aren't concrete follow up with questions to find out what is meant or intended. Does "good" mean beneficial, adequate, entertaining, or something else? For instance, if a housing program director suggests that the program is doing "okay," does he mean that the program is breaking even? meeting client needs? well liked by the community? at maximum use capacity? meeting or exceeding its goals? fully staffed? What exactly is "okay" about the program? In addition, remember that one person's "okay" is another's "good" and another's "less than we would like."

PLANNING, PRETESTING, AND PILOTING SURVEYS AND INTERVIEW PROTOCOLS

Relevant questions require systematic development. The first stage is developing questions. Before the questions are drafted the critical stakeholders should identify the information they want and agree that a survey, interview, or focus group will produce the desired information. As the questions are drafted you may want to meet with these stakeholders again to confirm that you are asking the "right questions." You should review the drafted survey and interview protocols with colleagues and ask them: Are any items ambiguous? Are the response choices appropriate? the information easily available to respondents? If their answers indicate that the questions are unambiguous, with appropriate response choices and accessible information, you have qualitative evidence that the questions are reliable. Writing questions takes time and effort; you need to have a thick skin. Good questionnaires and interview protocols may emerge only after criticism and argument. If you have already put considerable time and thought in developing the items, you may automatically disagree with colleagues' criticisms. In our experience, reviewers often identify the same items that actual subjects find troublesome.

After the questions are written and reviewed, the next step is to *pretest* the draft survey by asking a few people, similar to the prospective respondents, to answer it. For example, an employee survey should be given to a few employees ranging from unskilled laborers to senior management; they may also represent various divisions. A pretest normally involves face-to-face contact with the participants. You should record how long it takes individuals to answer the survey and if it seems to hold their interest. If a survey does not engage respondents, the information may be less accurate, incomplete questionnaires may be more common, or, in the case of mailed questionnaires, the response rate may be lower as respondents discard or ignore the survey.

Qualitative interview protocols also need to be pretested. Conduct interviews with several different people. Estimate how long the interview will take if a respondent answers all the questions. Note and revise questions that get respondents off topic. Document where you may need additional probes or should allow for silences. An interview that does not generate answers to your questions will leave you with lots of information but little that is relevant to your study. You will have wasted time, money, and resources.

At the end of the pretest consider asking the respondents for their reactions to the questions and the survey or interview as a whole. Also, use your own observations to identify potential problems. Are directions clear? Is the question order logical? Pretests commonly result in the investigators changing or clarifying unclear terms, rewording or dropping ambiguous questions, changing response categories, and shortening the survey or interview.

The last step is the *pilot study* or dress rehearsal. The study is implemented as planned on a small sample representing the target population. The planned analysis is carried out on the returned surveys and interview transcripts. The dress rehearsal should identify problems in contacting members of the sample and in getting their cooperation. The time involved in collecting and compiling the data can be estimated. The feasibility of the planned analysis can be verified. Previously unreliable, or invalid measures may be detected. Unused measures may be dropped. The investigators will especially want to note insensitive questions or questions that yield a large number of "Don't know" responses. Problems encountered in the pilot study should be resolved prior to implementing the survey.

In studies with a limited budget or a short timeline the pretest and the pilot study may overlap. At a minimum you should test the instrument or interview questions on potential subjects, check that items have sufficient variation, and conduct basic analysis to assess whether the survey will produce useful information.

Without conducting a pilot study and checking the proposed analysis, resources can be wasted. Consider an actual survey that resulted in 33,000 responses. The program staff spent a few frantic weeks simply compiling the responses to the 21 closed-ended and 2 open-ended questions; the total time spent just compiling the data was 550 hours. The data analysis was less costly, because the questions didn't allow for much analysis beyond summarizing the responses to each question. (Because of ambiguous wording all responses to two questions had to be discarded.) And remember, the time and cost investment would have been much greater had the data collection involved interviews. To conduct interviews and transcribe the data only to find that most of the conversations were off topic would be a devastating blow to any researcher!

CONCLUDING OBSERVATIONS

You may have begun reading this book assuming that quantitative research and qualitative research had little in common. We hope that by now you have observed that their sampling strategies, data-collection methods, and questioning techniques overlap. Both types of research require a defined research question that can be answered by the questions you ask and how you ask them. Both attempt to balance efficiency with obtaining quality information.

An oft repeated theme was the need to pretest questions and to obtain feedback from potential respondents and stakeholders. Another theme to keep in mind is making sure that questions don't encourage biased responses, socially desirable responses, and self-serving, and defensive responses.

RECOMMENDED RESOURCES

To keep up to date with current research on survey research topics such as question wording, questionnaire design, data collection, and data analysis, see *Public Opinion Quarterly,* the journal of the American Association for Public Opinion Research, published by the University of Chicago Press. Also see Recommended Resources listed at the end of Chapter 6.

Fowler, F. J., *Improving Survey Questions* (Thousand Oaks, CA: Sage Publications, Inc., 1995).

Fowler, F. J., *Survey Research Methods*, Fourth edition (Thousand Oaks, CA: Sage Publications, Inc., 2009).

Seidman, I., *Interviewing As Qualitative Research: A Guide for Researchers in Education and the Social Sciences* (New York: Teachers College Press, 2006).

CHAPTER 7 EXERCISES

There are three exercises for this chapter. The exercises develop your competence in writing and assessing questions and questionnaires.

- Exercise 7.1 The Long Street History Museum (Revisited) asks you to write questions for the Long Street History Museum study, to review them with a few classmates, and then identify in a class discussion qualities of good and bad questions.
- Exercise 7.2 Metro Citizen Kitchen and Food Pantry asks you to write questions and assess questions for the organization to use to routinely survey its clients.
- Exercise 7. 3 On Your Own asks you to apply a checklist to a questionnaire that you may have been asked to develop as part of an internship or your job.

EXERCISE 7.1 The Long Street History Museum (Revisited)

Scenario

In exercise 6.1 the director of the Long Street History Museum proposed two options to the board of directors so that the museum could continue to receive grants. The options were to (1) merge with another organization or (2) increase fund-raising efforts in order to operate without city grant money. Before doing so, however, the director wants to know what the museum members think about the options.

Section A: Getting Started

1. Write questions to ascertain the members' opinions about the two options.

Section B: Class and small group exercises

1. Work with one or two classmates to critique each other's questionnaires.
2. Participate in a class discussion where you
   a. identify common errors in question writing
   b. develop a list of examples of "good" questions and "poor" questions

EXERCISE 7.2 Metro Citizen Kitchen and Food Pantry

Scenario

Mohan Greene is the newly appointed executive director of the Metro Citizen Kitchen and Pantry (MECKP). The organization distributes food received from food drives, restaurants, and supermarkets to seven community soup kitchens and food pantries. Last year the kitchens served 13,000 meals, and pantries supplied food bags for 480,000 meals. (A food bag contains groceries adequate for three lunches or dinners.) Mohan is considering routinely surveying soup kitchen and food pantry users to see if the program is reaching its target clientele.

Section A: Getting Started

1. Write a closed-ended question that is
   a. a factual question
   b. a knowledge question
   c. a behavior question
   d. a motivation question
   e. an attitude or opinion question
2. For each question you wrote above indicate how Mohan might use the answers to improve MECKP.
3. Use a search engine to find a survey or survey questions for users of soup kitchens or food pantries. Use the information in this chapter and assess the introduction, question order, question, and response wording.
4. Use qualitative methods (see Chapter 2) and assess the reliability of the questions.
5. MECKP has a survey it plans to administer at the soup kitchens.
   a. Why should it do a pilot test?
   b. Write a list of tasks that need to be done as part of the pilot test.

Section B: Class simultation of drafting a questionnaire for MECKP

In groups of three to five draft a short survey and introduction for MECKP to administer to residents in its service area. It should have no more than 2 open-ended and 15 total questions. The questionnaire is to identify people's knowledge of what constitutes a good diet. MECKP plans to use the information to guide its development of an educational campaign.

1. Outline a short presentation that you would make to MECKP identifying what it may learn from the survey questions.
2. Make your presentation to the class and briefly review your questionnaire.
3. The class will select a questionnaire, make needed edits (improve question wording, response lists, and similar changes) and prepare for pretesting.
4. Each member of the class should administer the selected questionnaire to five people. Note how long it takes them to respond and any questions that they find confusing, unclear, or otherwise flawed.
5. Bring your completed questionnaires to class.
   a. Review each question and decide if it needs to be revised or deleted before including it in the final questionnaire.
   b. Compute the frequency of the responses to questions that you plan to use in the final questionnaire.
   c. Based on the frequencies does the class think that each question has sufficient variation to be worth keeping?

   Put the completed questionnaires aside for analysis after reading Chapter 8.

EXERCISE 7.3 On Your Own

If you have been tasked with developing a questionnaire, it may seem daunting. As discussed in this chapter, there are lots of considerations and choices to be made. Apply this checklist as you design your questionnaire. Once you are finished, review your questionnaire with your supervisor or other colleagues. Write a short memo assessing how well the checklist worked.

☐ Determine the purpose of your questionnaire.
    ☐ Learn what questions stakeholders want answered.
    ☐ Determine how stakeholders plan to use the answers.
☐ Determine the type of questions to ask.
    ☐ Do I need behavioral questions?
    ☐ Do I need factual questions?
    ☐ Do I need knowledge questions?
    ☐ Do I need opinion questions?
☐ My questionnaire has an introduction.
☐ My first questions are easily answered and clearly related to the survey's purpose.
☐ Questions are worded for optimal responses.
    ☐ Questions are worded so that a respondent understands what they mean.
    ☐ Questions are worded so that a respondent answers honestly.
    ☐ I have defined any unfamiliar terms.
    ☐ Questions specific about the time, place, or amount are included.
    ☐ Where appropriate, I have offered Don't know as a choice to discourage uninformed responses.
    ☐ Some statements go in different directions.
    ☐ Questions are stated so that they suggest to the respondent that any possible answer is acceptable
    ☐ Questions are free of bias.
    ☐ Loaded questions, which have words or phrases that evoke a strong positive or negative response, have been omitted.
☐ If I am using open-ended questions
    ☐ I am confident that respondents will be motivated enough to answer them.
    ☐ I will be able to manage the additional data compilation.
    ☐ I have the resources needed for the increased costs, specifically the cost of training staff to collect and compile the information and of verifying the reliability of summarized information.

Analyzing Survey Data
Describing Relationships Among Variables

After you have received completed surveys and forms and entered them in a database you must decide how to analyze them. You may start with the tools we covered in Chapter 5 and examine the frequency distribution of each variable. At a minimum this enables you to catch data entry errors and to identify insensitive measures, if any. In this chapter we cover tools you can use to examine relationships between two or more variables. We start with contingency tables, which effectively display relationships between variables measured at the ordinal or nominal level. We next cover linear regression, which examines the relationship among variables measured at the ratio or interval level. In connection with linear regression we also introduce correlation coefficients (r and R), measures of association that summarize the strength of a relationship. This chapter contains tools that can be appropriately used to analyze data from an entire population, a probability sample, or a nonprobability sample. Chapter 9 covers inferential statistics, which allow us to infer from a probability sample whether a relationship exists in the population.

ANALYZING NOMINAL AND ORDINAL VARIABLES: CONTINGENCY TABLES

You will find that exploring how variables are related is the most interesting part of research. After you first posed a research question you may have hypothesized how the values of one variable related to the values of another. You may wonder if the values of the independent variable are associated with the values of the dependent variable. For data measured at the ordinal or nominal level you may get the most information by organizing them in a contingency table. A *contingency table* shows the frequency or relative frequency of each value of the dependent variable for each value of the independent variable.

By convention, values of the independent variable head the columns and the values of the dependent variable head the rows. If the variables are ordinal, you should arrange the values along a continuum from low to high values. If the variables are nominal, however, no statistical consideration exists that recommends one arrangement over another. Instead, you can use your judgment to order the values. To avoid table clutter, you may want to include the number of respondents as part of the column label and report the relative frequencies, the percentages, in each cell so the column percentages sum to 100. If someone wants to know the number of cases in each cell, he can multiply the column total by the cell percentage. We illustrate this approach in Table 8.1, a contingency table relating respondents' opinion about local government service with their support for creating a fund to build affordable housing.

Managers and researchers use contingency tables to analyze relationships between variables. They may do this by identifying how attitudes, opinions, and support varyies among different groups. For example, affordable housing advocates wanted to know if public opinion about local government was related to support for a building fund. They were most interested in the difference between those who gave local government services high ratings and those who did not. They grouped opinions about local government services into three groups: "Excellent or Good," "Fair," and "Poor." The total in each group was treated as 100 percent, and the relative frequency of each group that responded with "For," "No opinion," and "Against" was computed. Each reader may take away different information from a contingency table. For example, write down what Table 8.1 shows you about the relationship between the two variables—"support for a fund to build affordable housing" and "opinion of local government service." Did you write "51 percent of the respondents favored the housing fund"? Or did you write that "55 percent of respondents who considered local services to be excellent or good supported the housing fund"? In both cases you focused on a single cell that dominates the rich information contained in the table as a whole.

To learn more from the table, you should focus on the percentages and the percentage differences between the values of the independent variable. Read across a row and note how the percentages change. Table 8.1 shows a direct relationship between support for creating the housing fund and opinion of local government services. The percentage of residents supporting the housing fund decreases as opinion of local government services decreases. The percentage of residents opposed to the housing fund increases as the opinion of local service decreases.

TABLE 8.1

Support for Creating Housing Fund by Opinion of Local Government Services

Support for Fund	Opinion of Local Government Services			
	Excellent or Good (%) ($n = 387$)	Fair (%) ($n = 224$)	Poor (%) ($n = 36$)	Total (%) ($N = 647$)
For	55	47	36	51
No opinion	24	25	34	25
Against	21	28	31	24

Note: Percentages may not add up to 100% due to rounding.

Another way to interpret the table is to focus on the center of gravity, represented by the median, for the different values of the independent variable. As the opinion of local government services becomes more negative, the median shifts further downward, away from support for the fund.

The percentage difference, the difference in percentages when subtracting across the rows, indicates the strength of relationship between two variables. The difference can range from 0, for no difference, to 100, indicating maximum difference. The approach is clearest when there are only two columns for the independent variable. Table 8.1 shows that those who gave highest ratings to government services were most likely to support the housing fund. For example, the percentage supporting the fund varies from 55 percent for those who gave government service a high rating to a low of 36 percent for those rating government services as "poor;" the percentage difference is 19.

Contingency tables are often referred to by size—the number of rows for the dependent variable and the number of columns for the independent variable. A two-by-four table has two rows for the dependent variable and four columns for the independent variable. Its title may list the dependent variable name first. For example, the title for Table 8.1 is Support for Housing Fund by Opinion of Local Government Services; it is a three-by-three table. Depending on the information you want you can calculate other percentages, such as the percentage of the total respondents in each cell, in which case you could say that "33 percent of all respondents both support the housing fund and gave local government services a high rating."

You may include a control variable to see if it impacts the relationship between the original two variables. For example, the housing fund advocates may wish to see if the relationship between service rating and support for the housing fund is the same for city residents and suburban residents. You would make residency the control variable, divide the original group of 647 cases into two groups—city residents and suburban residents—and create a table for each. The result would be Table 8.2.

Table 8.2 shows that the relationship between support for the housing fund and service rating is different for city residents and suburban residents. Those in the city were likely to be in favor of the housing fund if they rated government services as "Good" (65 percent), whereas suburban residents were unlikely to be for the fund even if they rated government services as Good (1 percent).

TABLE 8.2

Support for Housing Fund by Opinion of Local Government Services by Residence

Support for Fund	City (Service Rating)			Suburb (Service Rating)		
	($n = 328$) Good (%)	($n = 161$) Fair (%)	($n = 15$) Poor (%)	($n = 59$) Good (%)	($n = 63$) Fair (%)	($n = 21$) Poor (%)
For	65	46	30	1	50	35
No Opinion	20	25	35	45	25	34
Against	15	29	35	54	25	31

ANALYZING RATIO VARIABLES: LINEAR REGRESSION AND CORRELATION

If you have nominal or ordinal variables you may create contingency tables to examine relationships. The percents in the tables show how the values of one variable, the dependent variable, correspond to different values of the other variable, the independent variable. You can also create contingency tables for variables measured at the ratio level. Examples of ratio level variables include the number of hours of training, the number of volunteers who build houses for low income families, and the ages of people seeking help to locate safe, affordable housing. To create a contingency table for ratio variables, however, you may have to combine many values; otherwise you may end up with a large table that is virtually impossible to interpret. Because you will lose information if you combine values, linear regression and correlation may be better suited to analyze ratio level variables. Both *regression* and *correlation* are used across disciplines; they efficiently summarize data, handle a relatively large number of variables, and can be easily interpreted.

Consider an organization that builds houses for low income families and depends on volunteers to help on weekends. Knowing in advance how many volunteers will show up on any Saturday is important for scheduling and planning: materials, meals, and transportation all need to be arranged. The director of the organization suspects that the number of volunteers is related to how hot or cold it has been during the week. She asks a staff member to create a spreadsheet reporting the number of volunteers who showed up each weekend and the average daily high temperature for the week. Note that technically Fahrenheit temperature is an interval variable; however, you can treat variables measured at the interval level as ratio variables. Table 8.3 shows the spreadsheet data.

TABLE 8.3

Number of Building Volunteers and Average Daily High Temperature

Week	Average Temperature (°F)	Number of Volunteers
1	65	130
2	56	98
3	64	123
4	72	138
5	54	132
6	78	150
7	51	83
8	62	115
9	45	72
10	80	149

You should first graph the data to see if a relationship exists. The graph, called a *scatterplot*, plots the points that represent the values of temperature and the number of volunteers. You should display temperature, the independent variable, on the horizontal or *x* axis and number of volunteers, the dependent variable, on the vertical or *y* axis. You may think of the independent variable as the input variable, and the dependent variable as the output variable.

Note that the points on the scatterplot (Figure 8.1) fall along a line and as the average weekly high temperatures increase so do the numbers of volunteers. If the points can be enclosed in an imaginary envelope like a "thick cigar" then the relationship is said to be linear. The narrower the cigar, the stronger the relationship. The line in Figure 8.1 is called the regression line. Its formula is

$$Y = a + bX$$

where
Y = the value of the dependent variable;
X = the value of the independent variable;
a = the Y intercept or constant, the value of Y when X is zero;
b = the slope or regression coefficient.

The regression coefficient represents the change in Y for each unit increase in the value of X. The regression line, which can be calculated by statistical software, fits the data points better than any other straight line. Statisticians refer to this line as the ordinary least squares (OLS) line, because the sum of the squared differences

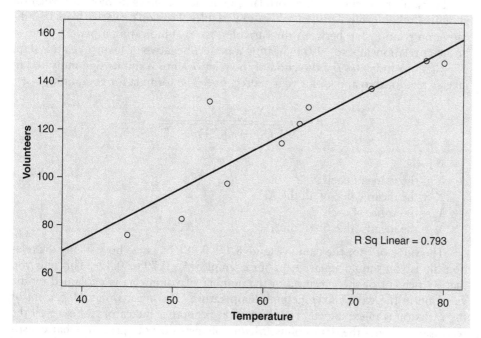

FIGURE 8.1
Number of Volunteers by Temperature

between each point and the regression line is smaller than the sum of the squared differences between each point and any other line.

The regression line in Figure 8.1 is $Y = -6.9 + 2X$. The director can use the regression equation $-6.9 + 2X$ to estimate how many volunteers will show up. If the weather forecast for the coming week is an average high of (60°F), the director could substitute 60 for X in the formula and calculate the value of Y, the number of volunteers. She would estimate 113 volunteers will show up [$y = -6.9 + 2.0 (60) = 113.1$] next weekend. Of course, the value of y is only an estimate: some weeks fewer volunteers than predicted will show up and other weeks more.

A regression line can be computed for any set of data. Your key concern is deciding how well the line describes or fits the data. The closer the points are to the regression line the better the fit. In Figure 8.1 the fit is quite good, although for one point ($x = 54°F$) the equation underestimated the number of volunteers by a large amount. The distance between a data point and the regression line for the same x value is called the residual. The larger the sum of the residuals, the poorer the fit. The poorer the fit, the less you can trust the estimate for Y.

You should not use the regression equation to estimate the value of Y for values of X that extend very far beyond the data used to calculate the regression equation. For example, the high temperatures used to calculate the regression equation for Figure 8.1 ranged from 40°F to 80°F. You should not assume that the line will give you an accurate estimate of how many volunteers will show up at the end of a very cold or a very hot week. If the temperatures edged toward 100—or higher—you should expect y to seriously overestimate the number of volunteers.

Pearson's r (or simply r), a correlation coefficient, gives us more information about the relationship between two ratio variables. Pearson's r varies from 0, indicating no relationship between the variables, to +1.0 indicating a perfect positive or direct relationship, to −1.0 indicating a perfect negative or inverse relationship. The sign of r indicates the direction of the relationship. A minus sign indicates an inverse relationship and a plus sign a direct one. The formula for computing r is:

$$r = \frac{\Sigma(X_i - \overline{X})(Y_i - \overline{Y})}{\sqrt{\Sigma(X_i - \overline{X})^2 \Sigma(Y_i - \overline{Y})^2}}$$

Where:
X_i = the value of each X
$\overline{X}$ = the mean value of all the Xs
Y_i = the value of each Y
$\overline{Y}$ = the mean value of all the Ys

The value of r for the data in Figure 8.1 is 0.89. To assess how strong the relationship is you should square r. In our example, r^2 is 0.79 or 0.89^2. This tells you that 79 percent of the variation in the number of volunteers is explained by the variation in the week's average high temperature. This means that 21 percent of the variation is unexplained. The value of r^2 is another indicator of how well the regression line fits the data points. Methodologists do not agree on what constitutes a weak, moderate, or strong relationship. We have heard different rules of thumb, such as an r of less than 0.25 describes a weak relationship and anything over 0.70 a strong one. We are somewhat skeptical of such rules of thumb. The strength of the relationship depends on the research question, findings from other studies, and how the data will be used.

Ideally you should use the scatterplot, regression equation, and the correlation coefficient together to assess the relationship between two interval variables. You should make it a practice to view the scatterplot; it will help you determine if the relationship is linear. If the relationship is not linear the correlation and regression statistics will not accurately portray the true relationship. You should be careful using regression with small data sets, since the regression line and correlation coefficient can be influenced by a single extreme value.

Multiple Regression: Regression and Correlation with More Than One Independent Variable

Regression is a powerful and versatile technique. *Multiple regression* expands the regression model to examine relationships with more than one independent variable. Although you may not need to use multiple regression in your own analysis, you are likely to run into it as you read reports and professional journals.

A multiple regression equation with two independent variables has the form

$$Y = a + b_1X_1 + b_2X_2$$

where
Y = the value of the dependent variable;
a = the intercept or constant, the value of Y when both independent variables are zero;
b_1 = the change in Y for a unit increase in X_1, controlling for X_2;
b_2 = the change in Y for a unit increase in X_2, controlling for X_1.

In a multiple regression equation each regression coefficient ($b_1 \ldots b_n$) controls for the effect of the other independent variables. That is, in the two-variable example, b_1 measures the change in Y if X_2 were held constant; b_2 measures the change in Y if X_1 were held constant.

Imagine that the director of the organization in our earlier example had asked a radio station to broadcast public service announcements (PSAs) soliciting volunteers, which the station did as its schedule allowed. The director's assistant consulted the station's records and added the number of weekly PSAs into the data set. For week (1) the data set showed 130 volunteers, an average high of 65°F, and 7 PSAs. The director used statistical software, which reported the following regression equation:

$$Y = -3.5 + 1.6X_1 + 3.7X_2$$

where
Y = the number of volunteers;
X_1 = the average high temperature;
X_2 = the number of PSAs.

The b_1 regression coefficient tells us that for every unit increase in average temperature, the number of volunteers increases by 1.6, controlling for the effect of the average number of PSAs. The b_2 regression coefficient tells us that each PSA increases the number of volunteers by 3.7, controlling for the effect of the temperature. Similar to the two-variable linear regression model, the resulting y value, the number of volunteers, is an estimate. The quality of the estimate depends on how well the multiple regression model fits the data.

R, the multiple correlation coefficient, is a multivariate measure of association and measures the strength of the joint relationship of independent variables and the dependent variable. We can judge how well the regression model fits the data by squaring R. R in this example is 0.93, so R^2 is 0.86. It indicates that the two variables explain 86 percent of the variation in the number of volunteers, an 8 percent improvement over just knowing the weekly high temperature. This R^2 value suggests that the multiple regression equation fits the data quite well. Knowing the value of two variables, weekly average high temperature and the number of PSAs explains 86 percent of the variation in number of volunteers.

If the independent variables are measured in different units or have widely different ranges, standardized regression coefficients facilitate comparing the independent variables. The standardized regression coefficients allow you to identify quickly which variables have the strongest relationships with the dependent variables and which have the weakest relationships. In our example temperature ranges from the low 40s to 80; the number of public service announcements ranges from 2 to 10. (The data on public service announcements and number of volunteers by week are shown in Table 8.5 for Exercise 8.2.) The standardized regression coefficient adjusts for different measurement units and different scales and tells us which independent variable has the most impact. The standardized coefficient for X_1 is 0.69 and for X_2 it is 0.33. The larger coefficient has the greater impact; in this example weekly average high temperature is more closely related to the number of volunteers who show up than PSAs are.

The standardized and unstandardized regression coefficients have different roles. The standardized regression coefficient allows you to directly compare the impact of the separate independent variables. The unstandardized regression coefficient allows you to estimate the value of y. To estimate the number of volunteers for a week that has an average high of 65°F and when the radio station ran 7 PSAs you would plug the respective temperature and PSA data into the equation with the unstandardized coefficients:

$$y = -3.5 + 1.6(65) + 3.7(7) = 126$$

In our examples the independent variables had a direct relationship with the dependent variable. Consequently, the values for the regression coefficients, the standardized regression coefficients, and r were positive. Had the relationship been inverse, the values would have been negative, but the strength would have stayed the same; that is, only the direction of the coefficients, not their absolute values, which indicate strength, would have been altered. For the purpose of illustrating this point let's imagine that the relationship between temperature and number of volunteers was $r = -0.89$ instead of +0.89. We would still say that we could explain 79 percent of the variation in number of volunteers. The change in our estimate would be that as temperatures go up the estimated number of volunteers would go down.

Regression describes a statistical relationship. When we say that a regression model fits the data, we mean that regression is an appropriate statistical model. However, a good fit does not necessarily mean that changes in the independent variable(s) *caused* changes in the dependent variable. To causally link variables we look beyond a statistical model to identify and eliminate other possible causes of an outcome. We discuss this in Chapter 11.

CONCLUDING OBSERVATIONS

In deciding how to analyze and present data you need to consider your research question, how your data are measured, and your audience. Contingency tables are appropriate for nominal and ordinal data. They also have the advantage of being easily understood. We have found that inexperienced researchers ignore the power of percentages: a clear, uncluttered table reporting relative frequencies is usually easier to interpret than one that only reports the raw numbers. After many years of teaching, we have learned that not all people interpret contingency

tables—even ones with relative frequencies—accurately. If you use contingency tables in your report, it's wise to include a sentence summarizing the findings.

Regression, especially multiple regression, is a powerful tool that can handle ratio variables efficiently. It has the distinct advantage of being able to include a number of independent variables in the same equation. Probably the greatest potential misuse of regression is to use it to analyze a small data set, where the addition of a few cases can greatly alter the findings.

RECOMMENDED RESOURCES

Arlene, Fink, *How To Conduct Surveys: A Step By Step Guide*. Fourth edition (Thousand Oaks, CA: Sage Publications, Inc., 2009). See especially Chapter 6.

Lee, E. S., *Analyzing Complex Survey Data*, Second Edition (Thousand Oaks, CA: Sage Publications, Inc., 2006).

O'Sullivan, E., G. Rassel, and M. Berner, *Research Methods for Public Administrators*, Fifth Edition (New York: Pearson/Longman, 2008). Chapter 13.

Rea, Louise, and Richard Parker, *Conducting and Designing Survey Research: A Comprehensive Guide*, Third Edition (San Francisco, CA: Jossey Bass, 2005).

Tufte, Edward R., *Data Analysis for Politics and Policy* (Englewood Cliffs, NJ: Prentice Hall, 1974) has a clear discussion with excellent examples on the use and interpretation of linear correlation and regression.

CHAPTER 8 EXERCISES

Analyzing Data to Find Relationships Exercises
    There are four sets of exercises for this chapter.

• Exercise 8.1 Nonprofit Participation in Experimental Financial Assistance Program is designed to give you practice in creating and interpreting contingency tables.
• Exercise 8.2 Do Public Service Announcements Yield More Volunteers? asks you to apply your knowledge of linear regression.
• Exercise 8.3 Physicians for Access also asks you to apply your knowledge of linear regression.
• Exercise 8.4 Fresh Start Center revisits the data presented in Table 5.8, and asks you to suggest strategies for identifying relationships.

EXERCISE 8.1   Nonprofit Participation in Experimental Financial Assistance Program

Scenario

One hundred (100) statewide nonprofits are asked to participate in an experimental program to deliver financial assistance. The local board of directors of each agency decides whether the organization will participate in the program. Table 8.4 contains the data on each nonprofit's region and whether it will participate in the program.

TABLE 8.4

Database on Participation in Experimental Program

Agency Number	Region of State	Experimental Program	Agency Number	Region of State	Experimental Program
1	Plain	No	51	Mountain	No
2	Plain	No	52	Mountain	No
3	Mountain	No	53	Coast	No
4	Plain	No	54	Mountain	Yes
5	Mountain	Yes	55	Plain	Yes
6	Mountain	No	56	Mountain	No
7	Coast	No	57	Plain	No
8	Coast	No	58	Plain	Yes
9	Coast	No	59	Coast	No
10	Coast	No	60	Coast	Yes
11	Coast	Yes	61	Coast	No
12	Mountain	No	62	Coast	No
13	Plain	Yes	63	Plain	Yes
14	Mountain	Yes	64	Coast	No
15	Coast	No	65	Coast	No
16	Coast	No	66	Coast	Yes
17	Plain	No	67	Coast	No
18	Coast	No	68	Plain	Yes
19	Plain	Yes	69	Coast	Yes
20	Mountain	No	70	Mountain	No
21	Coast	No	71	Plain	No
22	Mountain	No	72	Plain	No
23	Coast	No	73	Coast	Yes
24	Coast	No	74	Plain	No
25	Coast	Yes	75	Plain	No
26	Coast	No	76	Coast	No
27	Coast	No	77	Coast	No
28	Coast	No	78	Coast	No
29	Plain	No	79	Plain	No

TABLE 8.4

Database on Participation in Experimental Program

Agency Number	Region of State	Experimental Program	Agency Number	Region of State	Experimental Program
30	Coast	No	80	Plain	No
31	Coast	No	81	Plain	No
32	Plain	Yes	82	Mountain	No
33	Coast	No	83	Mountain	Yes
34	Coast	Yes	84	Coast	No
35	Plain	No	85	Plain	No
36	Coast	No	86	Plain	No
37	Coast	No	87	Plain	Yes
38	Mountain	No	88	Plain	No
39	Plain	No	89	Coast	No
40	Coast	No	90	Mountain	No
41	Plain	Yes	91	Coast	No
42	Coast	No	92	Mountain	Yes
43	Coast	No	93	Coast	No
44	Mountain	No	94	Plain	No
45	Mountain	No	95	Mountain	No
46	Coast	No	96	Coast	Yes
47	Coast	No	97	Coast	No
48	Coast	No	98	Plain	No
49	Plain	No	99	Coast	No
50	Mountain	Yes	100	Coast	No

Section A: Getting Started

1. Organize the data in a contingency table with three columns and two rows. Make Region of State the independent variable and Experimental Welfare Program the dependent variable.
   a. Tally the number of agencies in each region that are in the program and the number that are not. Report the number in each column. Fully label the table.
   b. In a second table, calculate the percentage of organizations in each region that are in the program and the percentage that are not. Enter only the percentages in the cells of the second table with the number of cases in each column at the head of the column. Fully label the table.
   c. What can you tell the state director about the relationship between the decision to participate in the Experimental Program and Region of the State?
2. What percentage of the agencies are in each region of the state?
3. What percentage of the agencies will participate in the experimental program? What percentage will not? What percentage of the agencies in the Plains region are participating?

4. Calculate and report the percentage difference between the percentage of agencies in the Mountains and the percentage in the Plains that are participating in the experimental program.

EXERCISE 8.2 Do Public Service Announcements Yield More Volunteers?

Scenario

A radio station runs public service announcements for an organization that has volunteers build houses for low-income families. The database representing the number of volunteers and the number of public service announcements is in Table 8.5.

Section A: Getting Started

1. Prepare a scatterplot to show the relationship between the two variables. Remember that "number of volunteers" is the dependent variable. Describe the direction of the relationship. Is it linear?
2. Compare your scatterplot for public service announcements to the scatterplot in Figure 8.1. How are they similar? How do they differ? Which relationship do you think is the stronger? Why?
3. The regression and correlation statistics for the relationship between number of volunteers and public service announcements are $a = 63.9$; $b = 8.5$; $r = 0.74$.
   a. Use this information to draw a regression line in your scatterplot.
   b. Use this information to write the regression equation and estimate the number of volunteers expected if the number of public service announcements is 4.

TABLE 8.5

Database on Number of Volunteers and Public Service Announcements

Week	Number of Volunteers	Public Service Announcements
1	130	7
2	98	2
3	123	6
4	138	10
5	132	8
6	150	9
7	83	5
8	115	6
9	76	5
10	149	7

EXERCISE 8.3   Physicians for Access

Scenario

Physicians for Access, a nonprofit organization, provides funds to increase the general population's access to medical care. To help in deciding how to distribute resources, the organization's director commissioned a study of the relationship between life expectancy and the number of people per physician in various areas. The organization assumes that life expectancy in a government jurisdiction is an indicator of general health conditions. Table 8.6 contains data from a sample of countries.

Section A: Getting Started

1. Create a scatterplot for the relationship between life expectancy and number of people per physician.
2. Describe the relationship shown in the scatterplot. Is it linear? positive or negative? weak, moderate, strong?
3. The regression and correlation statistics are $a = 72$; $b = -.002$; $r = -.61$.
   a. Use these statistics to further interpret the relationship.
   b. Write the regression equation relating the dependent variable, life expectancy, to the independent variable, number of people per physician.
   c. Calculate the life expectancy for a country with 450 people per physician.
   d. The actual life expectancy for the country in 3c is 76.5 years. Calculate the residual for this country.

TABLE 8.6

Database on Average Life Expectancy and Number of People per Physician

Country	Life Expectancy (Years)	Number of People Per Physician
A	71	370
B	65	684
C	70	640
D	78	400
E	57	2,470
F	77	233
G	61	7,610
H	56	2360
I	65	1,060
J	69	259
K	78	275
L	70	1,190
M	76	611
N	64	2,990
O	75	570

4. Suggest two additional independent variables that the researchers should consider for including in a multiple regression analysis of life expectancy.

EXERCISE 8.4 Fresh Start Center

Scenario

In Chapter 5 you were introduced to the Fresh Start Center, a community partnership that offers training in culinary arts, automotive repairs, and carpentry to unemployed workers. You have been asked to help the center develop a plan for monitoring its performance. Table 8.7 includes 25 cases from the first round of data collection. The data report the number of months of unemployment prior to entering training and the employment status of June 30 graduates as of August 1.

TABLE 8.7

Database on Fresh Start Participants

Id	Months Unemployed	Education	Skill Rating	Training Program	Status as of August 1
1	1	HS Grad	14	Culinary	FT Work
2	8	> HS	15	Automotive	Unemployed
3	7	> HS	14	Automotive	PT Work
4	11	> HS	14	Automotive	FT Work
5	1	HS Grad	12	Culinary	PT Work
6	8	> HS	13	Automotive	Unemployed
7	3	College Grad	12	Carpentry	Unemployed
8	11	College Grad	13	Automotive	Unemployed
9	5	< HS	13	Carpentry	FT Work
10	1	> HS	11	Carpentry	FT Work
11	4	HS Grad	14	Carpentry	PT Work
12	4	> HS	12	Carpentry	FT Work
13	9	HS Grad	14	Automotive	Unemployed
14	6	> HS	13	Automotive	FT Work
15	1	> HS	16	Culinary	Unemployed
16	1	> HS	13	Culinary	Unemployed
17	6	> HS	13	Carpentry	FT Work
18	6	> HS	10	Carpentry	Unemployed
19	5	> HS	12	Carpentry	Unemployed
20	4	< HS	12	Automotive	FT Work
21	6	< HS	14	Automotive	Unemployed
22	8	HS Grad	14	Automotive	Unemployed
23	8	HS Grad	12	Automotive	FT Work
24	1	> HS	15	Culinary	Unemployed
25	1	> HS	11	Culinary	FT Work

Section A: Getting Started

1. Create a contingency table to examine the relationship between training program and employment status.
   a. Write a sentence to describe the relationship shown in the table.
   b. Add education level as a control variable (you will want to combine values to create two groups). Does the control variable affect the original relationship? How?
2. Write a draft memo "Recommendations for Analyzing the Fresh Start Data." Remember, the actual analysis will be conducted on a large database. Your memo should indicate how you suggest analyzing the data to
   a. describe the clientele in Fresh Start's programs;
   b. describe how the clientele view what they learned in the program;
   c. identify the factors associated with the various outcomes.

Section B: Small Group and Class Exercises

1. Working with a group of two to four classmates develop a strategy for analyzing the database.
   a. Identify the tables that should be included.
   b. Use the database to construct an example illustrating what each table would look like.
   c. For each table explain the value of the information it would produce.
2. Review how each small group handled (i) describing the clientele, (ii) describing the clientele's perception of what they learned, (iii) factors associated with outcomes.
   a. Assess the value of each proposed table.
   b. Consider how the data can be organized most effectively.

APPENDIX A: Using Excel to Obtain Regression and Correlation Statistics

You can use Excel to calculate the correlation and regression statistics and to create a scatterplot. After entering the data in the spreadsheet

- click on the Tools menu;
- click on Data Analysis;
- click on Regression, then click OK. A dialogue box will appear.
  - In the box for Input Y range, enter the range of cells for the dependent variable.
  - In the box for Input X range, put in the range of cells containing the data for the independent variable. Click OK.

Try this with the data from Table 8.5 (in Exercise 8.2).

- In the Input Y range box, enter the cell range for the number of volunteers data.
- In the Input X range box, enter the cell range for the number of public service announcements data. Click OK.
- The output will include three boxes of statistics.

- First box—Multiple R. (It will not have a sign.) This is the value of r.
- The third box will contain the regression statistics.
- Look for the column headed Coefficients.
- The first value under the heading "Coefficients" is the value of a; to the left is its label "Intercept."
- In the second row, you will find the value of b; to the left is label "X variable."

You may need to determine the direction of the relationship from the sign of b, the regression coefficient, since Excel does not report a sign for the correlation coefficient. (This is another reason why viewing the scatterplot is important.) The b coefficient is also called the slope coefficient; it tells us the slope of the regression line. As noted earlier, the b coefficient shows the change in Y, the dependent variable, for a one unit increase in X, the independent variable. For a straight line that slope is either positive, as X increases y also increases, or negative, as X increases y decreases.

You can obtain a scatterplot by marking the two columns of data and clicking on the icon for "scatter".

Note: If the Data Analysis module is not available on the Tools menu, you will have to add it. For 1997–2003 versions of Excel, click on the Tools menu, then Add-Ins; then check Analysis Tool Pak. For the 2007 version click on Excel Options; on the drop down menu select Add-Ins. Click on GO at the bottom of the page. In the Add-Ins Box, check Analysis Tool Pak.

Generalizing from Survey Findings

Applying Inferential Statistics

You may be so used to hearing what a poll found that you rarely stop and think about how a sample of roughly 1,000 people can accurately represent the country's population. Similarly, you may have heard reports that researchers analyzing survey data found that "people who drink moderate amounts of red wine have fewer heart attacks" or that "senior citizens who play bridge are less likely to suffer from dementia." In presenting the findings from sample data, pollsters and researchers rely on inferential statistics. The statistics we have presented thus far are descriptive statistics, and you can use them to summarize and describe any set of data. Inferential statistics estimate parameters and indicate if a relationship probably occurred by chance. To use inferential statistics correctly you must have data from a probability sample. Otherwise, you have to rely on descriptive statistics to present your findings and make decisions.

In this chapter we cover sampling statistics and tests of statistical significance. Sampling statistics guide decisions that allow us to infer who will win an election and how Americans feel about the economy. Next we discuss tests of significance, which help us decide if a relationship that exists in data from a sample probably occurred by chance. Tests of statistical significance cannot tell us that red wine will decrease heart attacks or that playing bridge will prevent dementia. Tests of significance simply eliminate chance as an explanation for the relationships; however, the statistical findings may stimulate more rigorous research to explain why a relationship exists and to eliminate alternative explanations.

SAMPLING STATISTICS

You should have a basic knowledge of sampling statistics to determine sample size and to interpret findings from a probability sample. To apply and interpret sampling statistics correctly you need to understand the terms *parameter*, *sampling error*, *standard error*, *confidence interval*, and *confidence level*. A parameter is a characteristic of the population you are studying. An example of a parameter is the percentage of all food pantry users who are unemployed. Statistics refer to a characteristic of a sample, such as the number of sampled pantry users who are unemployed. With statistics from probability samples, you can estimate parameters. We are sure that you do not expect a sample of 400, 1000, or any other size to indicate the exact percentage of unemployed. *Sampling error* refers to the difference between the parameter and a statistic. You can use the sampling error to mathematically estimate a parameter. The sampling error is based on sample size and the sample's standard deviation, which estimates the population's variability.

The standard error is the standard deviation of a theoretical distribution of values of a variable. Think of it this way, if you drew an infinite number of random samples of the same size, 95 percent of them would yield a statistic, such as a mean, that is within ±1.96 standard errors from the parameter. Although we can't draw an infinite number of samples we know that 95 out of a 100 random samples will fall within ±1.96 standard errors of the parameter; 5 out of 100 of the samples will fall outside this interval.

The confidence level is the probability that the parameter will fall within a given range. With a 95 percent confidence level, the level commonly used in social science research, you know that in 95 random samples out of 100 a parameter will fall within ±1.96 standard errors of a statistic. We refer to this characteristic as the 95 percent confidence level. Occasionally you may see a 99 percent confidence level, which would be ±2.58 standard errors. At a 99 percent confidence level, the parameter is less likely to fall outside the confidence interval, but the confidence interval is wider.

Confidence intervals identify the range where the parameter probably falls. To determine the confidence interval you have to know the value of the standard error. Let's imagine sampling food pantry users to estimate the percentage unemployed. If you drew a large number of probability samples of the same size and graphed the percentage unemployed for each sample, the graph would take on a bell shape. Most of the samples should cluster near its center. A few samples, however, will be located at the far ends of the graph. These samples indicate a much lower percentage of unemployed or a markedly higher percentage of unemployed. From looking at the graph you would feel confident in estimating that the parameter, the actual percentage unemployed, is somewhere in the interval where the samples are clustered. The interval is the *confidence interval*.

Let's use an example to review our discussion and to help you understand the various terms and what they tell you.

> *Problem:* An agency needs to report the average earnings of participants in a job training program after they complete the program.
>
> *Population:* All participants who have completed the job training program within the past 2 years.
>
> *Sampling strategy:* Construct and contact a probability sample of 100 participants and ask for each person's current annual salary.
>
> *Finding:* The annual average (mean) salary of the sampled participants is $25,000. The standard error, estimated from sample data, is $250.
>
> *Interpretation:* The agency can be 95 percent confident that the average salary of all recent training graduates is between $24,510 and $25,490 (1.96 standard

errors below the mean and 1.96 standard errors above the mean). In other words, the confidence interval is between $24,510 and $25,490. There is a 5 percent probability that the parameter lies outside the confidence interval.

Explanation: If all possible random samples each consisting of 100 job training participants were drawn, the average salaries found in 95 percent of the samples would fall between 1.96 standard errors below and 1.96 standard errors above the parameter, which is the true average. We do not know if any given sample falls within this confidence interval.

Pollsters typically refer to the sampling error when they report proportional data, that is, data reporting the percentage of cases for each value or category. By convention the sampling error equals the standard error at the 95 percent confidence level. For example, if 60 percent of the sample had finished high school and the sampling error is 3 percent (1.96 times the standard error of 1.5 percent), you could report that you are 95 percent confident that between 57 and 63 percent of all trainees completed high school. Since you are most likely to work with proportional data, let's examine the accuracy of the statement, "The results from the full survey of 1,000 randomly selected adults have a margin of sampling error of plus or minus 3 percentage points."

Let's start by calculating the sampling error for a 95 percent confidence level. The equation is

$$\text{Sampling error} = \sqrt{\frac{p(1 - p)}{n}} \times 1.96$$

where
p = the proportion in a given category;
n = the sample size.

The largest possible sampling error assumes maximum variability, that is, a 50-50 split. So if $p = 0.5$, $1 - p = 0.5$, and $n = 1,000$, the sampling error is 3.1 percent. Depending on how you will use the data, you can decide whether you estimate the sampling error for all the proportional findings using a 50-50 split or if you should calculate the error for each finding. If we used the sample's finding of 60 percent high school graduates as the value of p, the sampling error would be 3 percent. For nominal and ordinal variables with more than two categories you can treat the values as dichotomies. For example, assume that the sample of food pantry users include full-time workers, part-time workers, unemployed workers, and retirees. To estimate the parameter for the percentage unemployed, have p equal the percentage unemployed in the sample, and $(1 - p)$ equal the percentage in all other categories.

The sampling error of 3 percent applies to the entire sample. It does not apply to a specific group within the sample. Let's assume that you have a sample of 500 food pantry users. Let's also assume that 75 percent of them are high school graduates. To estimate the percentage of pantry users who are high school graduates you would decide on one of the following equations to calculate the sampling error.

$$\text{Equation 1: Sampling error} = \sqrt{\frac{.5(.5)}{500}} \times 1.96 = 4.4\%$$

$$\text{Equation 2: Sampling error} = \sqrt{.75\left(\frac{.25}{500}\right)} \times 1.96 = 3.8\%$$

Equation 1 assumes maximum variability and gives us the largest sampling error for a sample of 500. You may prefer this equation if you are examining several characteristics of food pantry users and a rough estimate is adequate. This is often the case. On the other hand if you need a more precise estimate you may prefer Equation 2. In both cases there is a 5 percent probability that your estimate is wrong and the parameter is outside the confidence interval.

Let's step back and use a simple example to underscore some key points. A study found that 86 percent of the 245 working mothers surveyed reported feeling stress.[1] For a 95 percent confidence level the sampling error is 4.4 percent. Using the sampling error, you would estimate that between 81.6 and 90.4 percent of working mothers feel stress. The 86 percent is the statistic. You estimated that the parameter, the actual percentage of stressed working moms, falls within the confidence interval, that is, between 81.6 and 90.4 percent. With a 95 percent confidence level there is a 5 percent chance that your estimate is wrong; 5 times out of 100 a sample of 245 will either underestimate or overestimate the location of the parameter.

Sample Size

A characteristic of samples that you may wish to remember is that the larger the sample the smaller the sampling error and vice versa. To decide on the sample size, you need to decide how accurate you want the sample to be. Of course, practical considerations such as the amount of time and money available may limit sample size. The following principles highlight how accuracy, confidence level, and population variability guide decisions about sample size.

Accuracy: The greater the accuracy desired, the larger the sample needs to be.

Confidence level: The more confidence desired, the larger the sample needs to be.

Population variability: The more diverse the population, the larger the sample needs to be.

Although the accuracy improves with larger samples, the amount of improvement may become less and less. It is a classic case of diminishing returns. When the sample size is small—say 100—increasing it to 400 will greatly improve its accuracy. However, an increase from 2,000 to 2,300 will bring little improvement, although the additional costs of adding 300 respondents is likely to be the same in both cases. For this reason, a sampling error of 3 to 5 percent and a 95 percent confidence level are common in social science research.

---

[1]Parker, Kim, "The harried life of the working mothers," *Social and Demographic Trends*, Pew Research Center, October 1, 2009. Posted at http://pewsocialtrends.org/pubs/745/the-harried-life-of-the-working-mother#prc-jump. Accessed January 27, 2010.

To illustrate how to compute the sample size, assume we want to see what percentage of residents has used a food pantry within the past year. We have decided to accept a sampling error of 4 percent. The equation to find the sample size for proportional data is

$$n = p(1 - p)\left(\frac{1.96}{\text{sampling error}}\right)^2$$

Just as we did with estimating the sampling error we will assume maximum variability, that is, 50 percent in one category and 50 percent in the other.

$$n = .5(.5)\left(\frac{1.96}{.04}\right)^2$$
$$= .25 \times 2{,}401$$
$$= 600$$

If you have little or no information about the population, the easiest and most conservative approach is to assume that the population is split 50-50. In the above example 600 residents is the maximum sample needed for a 4 percent sampling error. If you are familiar with the population you may prefer to make a less conservative estimate of the variability. If you estimated that no more than 30 percent of the population has used a food pantry your calculation to determine sample size with a 4 percent sampling error would be

$$n = (.3)(.7)\left(\frac{1.96}{.04}\right)^2$$
$$= .21 \times 2{,}401$$
$$= 504$$

Another important factor in determining sample size is how much analysis you plan to do. Small samples will not withstand extensive analysis. With a small sample and several independent or control variables, you can find yourself studying individual cases—nobody wants to generalize from one or two cases. Similarly, as you study groups within a sample, subsamples will be smaller. For example, a statewide sample of 600 may be inadequate for examining differences between counties.

TESTS OF STATISTICAL SIGNIFICANCE

Students often find tests of statistical significance difficult to grasp the first time they encounter them. The word *significance* implies that the tests are very important. In a sense they are, yet the information they provide is modest. Let's begin by imagining that investigators randomly selected 100 people at a rock concert, took their blood pressure, and asked questions about their lifestyle. (We will leave you to imagine the logistics required to sample and collect data from attendees at a rock concert.) If the wine drinkers had lower blood pressure, you need to ask, "Is the relationship between drinking wine and blood pressure a coincidence, and if we collected data from the entire population of concert attendees, would we have found the same relationship?"

Using the language of researchers we might ask if wine drinking and blood pressure are *independent* of each other or if the two variables are *randomly related* (same as not related). Asking if variables are randomly related or independent of one another is the same as asking "Could this relationship have occurred by chance?" If a statistical test suggests that in the population—the rock concert attendees—the relationship between wine drinking and blood pressure is nonrandom, the relationship is said to be *statistically significant.*

At the core of statistical significance is hypothesis testing, which has its own terminology. The terminology, which is what students often find confusing, reflects careful statistical and epistemological thinking underlying hypothesis testing.

Before we go into more detail we should let you know that respected social scientists consider statistical significance overrated, misunderstood, and frequently misused.[2] We agree. Nevertheless, we cannot in good conscience completely ignore the topic: The word *significance,* the time spent on hypothesis testing in statistics courses, and the frequent appearances of significance tests in research reports all suggest that tests of statistical significance are extremely important. But, a statistically significant relationship may not be strong, important, or valuable. Large samples may show that trivial relationships are statistically significant. A finding of statistical significance does not tell us that the research was conducted correctly. The measures may have been unreliable or the data may be from a nonprobability sample.

Our goal is to help you correctly interpret what a test of statistical significance tells you, and what it doesn't tell you. As part of our discussion we include the calculations for two common tests. Although you may never do the calculations yourself they may help you better understand the tests. Even though our discussion may seem terminology-heavy we will focus only on the most relevant terms.

The process for determining if two variables have a nonrandom relationship in the population has four steps:

1. State the null and alternative hypotheses.
2. Select an alpha level, that is, the amount of risk you are willing to accept that you are wrong if you reject the null hypothesis.
3. Select and compute a test statistic.
4. Make a decision.

Stating the Null Hypothesis

A *hypothesis* states a relationship between two variables. The *null hypothesis* (H_0) postulates no relationship, or a random relationship, between the same two variables. The alternative hypothesis (H_A), also called the research hypothesis,

---

[2]Cohen, J., "Things I have learned (so far)," *American Psychologist* (1990), 45:1304–1312 provides an accessible discussion of the limitations of tests of significance. Gill, J., "The insignificance of null hypothesis significance testing," *Political Research Quarterly* (1999), 52:647–674 and Henson, R. K., "Book Review: Beyond significance testing: Reforming data analysis methods in behavioral research," *Applied Psychological Measurement* (2006), 30:452–455. Both summarize the basic arguments and give extensive references. We have avoided addressing the arguments because both our position and theirs should lead you to avoid overvaluing the results and to instead focus on the strength of relationships and the quality of the research methodology.

postulates a relationship between the variables. Let's go back to our blood pressure study at the rock concert. Had the investigators examined the relationships between blood pressure and other variables such as exercise, diet, and medications, some relationships would have been strong, others weak. The first question they would want to ask is, "What is the probability we found a relationship by chance?" To answer this question they state the null hypothesis and the alternative hypothesis.

H_0: There is not a relationship between drinking wine and blood pressure.

H_A: The blood pressure of wine drinkers is lower than that of non–wine drinkers.

Remember the subjects represent only one sample. You must allow for the possibility that sampling error may account for the lower blood pressure of wine drinkers. Different samples may yield different results. Another sample from the same population may show no relationship between drinking wine and blood pressure.

To decide if the null hypothesis is probably true in the population, you should use a test of statistical significance. If the test suggests that the null hypothesis is probably untrue, you would reject your null hypothesis and accept the alternate hypothesis. In doing so, you risk making a mistake and accepting an untrue alternative hypothesis. Rejecting a true null hypothesis is called a *Type I* error. A Type I error occurs if, based on the sample, you decide that the alternative hypothesis is true in the population, when in fact it is untrue. A Type I error may be thought of as a false alarm; in other words it calls attention to a relationship that does not exist.

Alternatively, you may fail to reject an untrue null hypothesis and assume that a relationship does not exist in the population when in fact it does. In making this decision your mistake is to discount a true alternative hypothesis. Failing to reject an untrue null hypothesis is called a *Type II* error: based on the sample, you decided that the alternative hypothesis was untrue, when in fact it was true. The following illustrates our discussion of the two types of error.

Type I error: Based on sample data we reject the null hypothesis and accept the alternative hypothesis that drinking wine decreases blood pressure; in reality drinking wine may not be related to a decrease in blood pressure.

Type II error: Based on sample data we fail to reject the null hypothesis and assume that drinking wine is not related to blood pressure when in reality it is.

The bottom line is that a test of statistical significance helps you to decide whether your alternative hypothesis is true or untrue in the population. Based on your initial analysis, you already know if it is true or not in your sample. We should note that all a test of significance does is to say that the observed relationships probably did not occur by chance. It does not tell you if the relationship is stronger, weaker, or the same in the population. Nor does it prove that wine drinking will lower blood pressure. Factors other than drinking wine could cause the lower blood pressure. All the test of significance does is to help eliminate chance as one of the factors in accepting a hypothesis.

Selecting an Alpha Level

Traditionally, researchers select a criterion for rejecting the null hypothesis prior to starting their analysis. This criterion, referred to as the *alpha* (α) *level,* reports the probability that the variables are unrelated in the population. The alpha level is a number between 0 and 1. Common alpha levels for hypothesis testing are 0.05, 0.01, and 0.001.

$\alpha = 0.05$ indicates a 5% chance of committing a Type I error

$\alpha = 0.01$ indicates a 1% chance of committing a Type I error

$\alpha = 0.001$ indicates a 0.1% chance of committing a Type I error

If you decrease the probability of a Type I error you increase the probability of a Type II error. If you change α from 0.05 to 0.01 your chance of a Type I error goes from 5 percent to 1 percent, but at the same time the probability of missing a true alternative hypothesis (Type II error) increases. The only way to decrease both types of error at the same time is to increase the sample size.

In selecting an alpha level you should consider the sample size, the strength of the relationship, and the practical consequences of committing a Type I error or a Type II error. If alpha is set at 0.05 a sample of 1,300 will detect a slight difference or effect 95 percent of the time, and a sample of 50 will detect a moderate effect only 46 percent of the time.[3] Tables 9.1a, 9.1b, and 9.1c are three hypothetical tables linking wine drinking and blood pressure. Given 0.05 as the alpha level, the relationship between the variables in Tables 9.1a and 9.1c is statistically significant; the one in Table 9.1b is not.

TABLE 9.1A

Blood Pressure by Wine Drinking

Blood Pressure	Number of Respondents That Drink Wine	Number of Respondents That Do Not Drink Wine
High	20	22
Normal or low	40	18

TABLE 9.1B

Blood Pressure by Wine Drinking

Blood Pressure	Number of Respondents That Drink Wine	Number of Respondents That Do Not Drink Wine
High	20	20
Normal or low	40	20

[3]Cohen, J., "Things I Have Learned (So Far)," (p. 1308) This discussion is based on the concept of power, the probability of correctly rejecting a null hypothesis. To determine a study's power one needs to have specified the sample size, alpha level, and effect size. We believe that our discussion, which underscores the role sample size and effect size, is adequate for most readers.

TABLE 9.1C

Blood Pressure by Wine Drinking

Blood Pressure	Number of Respondents That Drink Wine	Number of Respondents That Do Not Drink Wine
High	100	100
Normal or low	200	100

How do these relationships differ? The sample in Table 9.1c is five times larger than that in Table 9.1b. Table 9.1a shows a stronger relationship than Table 9.1b. All three tables would be statistically significant if we had selected 0.10 as the alpha level. The following summarizes the factors that affect whether a relationship is found to be statistically significant.

The larger the sample the more likely you are to find statistical significance.

The smaller the sample the less likely you are to find statistical significance.

The stronger the relationship the more likely you are to find statistical significance.

The higher the alpha level the more likely you are to find statistical significance.

The lower the alpha level the less likely you are to find statistical significance.

Selecting and Computing a Test Statistic

In this text we examine two common tests of statistical significance: *chi-square* (χ^2), a statistic for nominal level data usually applied to contingency tables[4] and the *t-test*, a statistic that compares the differences between the means of two groups.

Chi-square: A chi-square test compares the frequencies in a contingency table with the frequencies that would be expected if the relationship between variables is random in the population. Table 9.2a tests a hypothesis that the job training

TABLE 9.2A

Frequencies Observed: Outcomes by Type of Training Program

	Vocational Education	On-the-Job Training	Work Skills Training	Total
Working	19	109	164	292
In School	19	82	31	132
Unemployed	26	82	54	162
Total	64	273	249	586

---

[4]Chi-square can also be used to determine goodness of fit; here, we have limited our discussion to the use of chi-square to tests for independence between variables.

TABLE 9.2B

Frequencies Expected: Outcomes if Type of Training Program Has No Effect

	Vocational Education	On-the-Job Training	Work Skills Training	Total
Working	31.9	136.0	124.1	292
In School	14.4	61.5	56.1	132
Unemployed	17.7	75.5	68.8	162
Total	64	273	249	586

programs have different outcomes. The table compares participants in three job training programs and their outcomes. Each cell contains the frequencies observed (f_o) in the collected data.

Table 9.2b shows what the data would look like if there was no relationship between a program and trainee outcomes; each cell contains the frequencies expected (f_e) if the null hypothesis were true.

The frequency, f_e, for each cell is the column total multiplied by the row total divided by the sample size. For example $31.9 = (64 \times 292)/586$.

Typically software programs calculate chi-square using the equation $\chi^2 = \Sigma[(f_o - f_e)^2/f_e]$ and report the value of p, the associated probability. The associated probability is the actual level at which the value of a statistical test is significant. The associated probability may be lower, higher, or the same as alpha. In this example $\chi^2 = 50.67$. The associated probability may be reported as 0.0000. There is less than 1 chance in 1,000 that you would obtain χ^2 equal to 50.67 if the variables were independent, that is, not related. You might report your finding in a table footnote, "$\chi^2 = 50.67. \ p < 0.001$," or in a sentence, "The type of job training is related to what trainees are currently doing $(\chi^2 = 50.57, p < 0.001)$." The important piece of information is "$p < 0.001$," which provides strong evidence that the relationship is probably not random. Citing the statistic (χ^2) allows readers trained in statistics to determine whether you used the appropriate statistical test. Some researchers also report degrees of freedom (df), which enables reviewers to visualize how the data were analyzed.[5]

Chi-square has two characteristics that you want to keep in mind. First, as a nominal statistic, it does not provide information on the direction of any relationship. In our example, chi-square indicates that the relationship between type of training program attended and current status is probably nonrandom. It does not indicate which program is the most effective. Second, the numerical value of chi-square tends to increase as the sample size increases. Thus the chi-square value is partially a product of sample size. It does not directly measure the strength of the association between variables and should not be used as a measure of association.

---

[5]The value of most tests of statistical significance is affected by sample size, degrees of freedom, or both. For chi-square the degrees of freedom (df) are based on the number of rows and columns in the table.

t-tests: The *t*-test is a statistic for ratio data. It tests hypotheses that compare the means of two groups. You can test hypotheses that one group's mean is higher than another group's, in which case you use a one-tailed test. Alternatively, you can test hypotheses that the group means are different, in which case you use a two-tailed test. In our example, if we measured actual blood pressure the hypothesis that wine drinkers have lower average blood pressure than non–wine drinkers would require a one-tailed test. The following example hypotheses illustrate two- and one-tailed *t*-tests.

H_A: The average salary of male trainees and female trainees differs (use a two-tailed test).

H_0:   The average salary of male and female trainees is the same.

H_A: The average salary of male trainees is higher than the average salary of female trainees (use a one-tailed test).

H_0:   The average salary of male trainees is the same as or less than the average salary of female trainees.

A two-tailed test does not specify whether men or women earn more. If you found that the average salary of female trainees is greater than that of male trainees, you would reject the null hypothesis. In a one-tailed test, the null hypothesis is expanded to include a finding in the "wrong" direction. Thus, if the *t*-test implied that the average salary of female trainees was more than that of male trainees, you would not reject the null hypothesis.

For *t*-tests comparing two groups you can assume either equal or unequal variation in the population; the population variation is estimated by the standard deviation. The equation for unequal variation can serve as a default option; it is more conservative and produces slightly higher associated probabilities. To illustrate a *t*-test, we assume unequal variations of male and female's salaries.

For the Male Sample	For the Female Sample
n1 = 403	n2 = 132
Mean1 = $17,095	Mean2 = $14,885
Standard deviation(s)1 = $6,329	Standard deviation (s)2 = $4,676

A software program may calculate the value of *t* using the equation

$$t = \frac{\overline{X}_1 - \overline{X}_2}{\sqrt{\left(\frac{s_1^2}{n_1 - 1} + \frac{s_2^2}{n_2 - 2}\right)}}$$

and report its *significance level*. In this example, $t = 4.28$. Its associated probability may be reported as 0.0000. Reports normally include the value of *t*. (If *t* is at least 2, then $p \leq 0.05$ for either a one-tailed or a two-tailed test.) You may report the findings in a sentence, "The average salary of male trainees is higher than that of female trainees ($t = 4.28, p < 0.001$)," in a table with the value of *t* in one column and the associated probability in the next, or in a table footnote. In the case of table footnotes a common practice is to place asterisks next to the *t* values and give the alpha level in a table footnote. The relationship between the number of asterisks and value of *p* varies from author to author. Usually, the more asterisks, the lower the value of p (you will see an example of this in Exercise 9.4). Students who have studied statistics may wonder why we do not use the normal distribution (z-scores) to test hypotheses about sample means. To use z-scores properly, the population variance must be known. If the population variance is not known, it is estimated by the standard deviation, in which case t-tests as opposed to z-scores are appropriate. Therefore, while using z-scores introduces little error with larger samples ($n > 60$), social scientists tend to rely on t-tests.

Making a Decision

If your significance statistic yields a low associated probability you can assume that the null hypothesis of no relationship is untrue. In our examples you would reject the null hypotheses and accept the alternative hypotheses that different job-training programs achieved different outcomes and that male trainees earned more than female trainees. The associated probability does not indicate the probability of a Type I error, nor does it imply that a relationship is more significant or stronger. Its contribution is more modest. It indicates the probability of a specific χ^2 or t-value occurring if the null hypothesis is true. The relationship may be weaker, or stronger, than the one found in the sample data.

The findings result from many decisions, including whom to sample, how to sample them, what information to gather, how to gather it, and when to gather it. Remember that you are applying a statistical test to a set of numbers. Statistical significance cannot make up for a flawed design. You will obtain an answer even if your sample was biased and your measures unreliable. Even if the methodological decisions are sound and the statistics are applied correctly, the data still represent a single sample out of all the possible samples from the population. A test of significance alone should not bear the burden of demonstrating the truth of a hypothesis. It is far more realistic and reasonable to consider each statistical finding as part of a body of evidence supporting the truth or error of a hypothesis.

If you fail to reject your null hypothesis you do not have irrefutable evidence that the null hypothesis is true—a different sample or a larger sample might yield different results.

ALTERNATIVES TO TESTS OF STATISTICAL SIGNIFICANCE

Social scientists have recommended four alternatives to the traditional significance test. First, the null hypothesis can include a specific difference. Let's go back to an earlier example. We will assume that administrators do not want to fund more on-the-job training programs unless they place at least 10 percent more of their trainees than other training programs. To test this preference the null hypothesis and the alternative hypothesis would be stated as

H_A: On-the-job training programs place at least 10 percent more of their trainees than other training programs.

H_0:   On-the-job training programs do not place at least 10 percent more of their trainees than other training programs.

Second, you should consider reporting confidence intervals. Proponents argue that confidence intervals are more informative than a test of significance. They provide information on the value of parameters, differences between them, and the direction of the differences. The confidence interval avoids implying that the sample means, for example, provide a precise estimate of the differences. Rather the population means probably fall somewhere within the range indicated by the confidence intervals.

Third, you may include the measure of association. Statistics such as r and r^2, not tests of statistical significance, measure the strength of a relationship.

The fourth alternative is replicating studies. The importance of replication is underscored by the following quote, "the results from an unreplicated study, no matter how statistically significant . . . are necessarily speculative. . . . Replications play a vital role in safeguarding the empirical literature from contamination from specious results."[6] Replications do not have to duplicate previous research exactly. Investigators may implement the research using a different population or another setting and see if the findings generalize to other populations or settings. Findings that go in the same direction, whether or not they are statistically significant, or that have overlapping confidence levels provide more support than a simple significance test.

With the exception of measuring the size of the effect of the independent variable on the dependent variable, each alternative has constraints. You may have inadequate knowledge to specify a relationship beyond anticipating some difference. Confidence levels work with interval data, but setting confidence levels for proportional data are less informative. Replication requires the opportunity and resources to repeat a study.

CONCLUDING OBSERVATIONS

Sampling statistics and tests of statistical significance are inferential statistics, that is, we can infer something about our population from a probability sample. Sampling statistics allow us to estimate the value of a population characteristic, the parameter. Tests of statistical significance, when applied to relationships between variables, simply indicate the probability that a null relationship exists in the population even if the data show a relationship. Sampling statistics are straightforward and less subject to misinterpretation. Consider the statement "a survey of River Dale residents, which had a 6 percent sampling error, found that 36 percent of the respondents used a food pantry last year." We would estimate that between 30 and 42 percent of River Dale residents used a food pantry last year. Because the surveyors used a 95 percent confidence level, which is almost always used in social science research, 5 times out of a 100, the characteristic will lie outside the 30 to 42 percent range.

Tests of statistical significance are trickier. The term *significance* may imply that a relationship is important or strong. One may erroneously assume that a strong sample relationship indicates a strong relationship in the population or that an intervention caused the observed outcome. None of these are true. The relationship may be unimportant or weak in the population. The intervention may not have caused the outcome. The current trend is to put less emphasis on a single statistic. Instead researchers report confidence intervals and effect sizes (such as r^2).

---

[6]Hubbard, R., and Ryan, P. A., "The historical growth of statistical significance testing in psychology— And its future prospects," *Educational and Psychological Measurement* (2000), 60:661–681.

Both tests of statistical significance and sampling statistics may give a false impression of certainty. Neither can compensate for invalid or unreliable measures or sloppy data collection. Even properly drawn samples are subject to error. However, tests of significance do have value in that they provide evidence to interpret research findings.

RECOMMENDED RESOURCES

Fink, Arlene, *How To Conduct Surveys: A Step By Step Guide*, Fourth Edition (Thousand Oaks, CA: Sage Publications, Inc., 2009). See especially Chapter 6.

O'Sullivan, E., G. Rassel, and M. Berner, *Research Methods for Public Administrators*, Fifth Edition (New York: Pearson/Longman, 2008). Chapter 12.

Rumsey, D, *Statistics for Dummies* (Hoboken, NJ: Wiley Publishing, Inc., 2009).

Salkind, Neal, *Statistics for People Who (Think They) Hate Statistics*, Fourth Edition (Thousand Oaks, CA: Sage Publications, Inc., 2010).

CHAPTER 9   EXERCISES

There are four sets of exercises for Chapter 9.

- Exercise 9.1 Reviewing Polling Data reports a poll on racial discrimination. This exercise is designed to give you practice in computing and interpreting sampling statistics.
- Exercise 9.2 Attitudes toward Corporal Punishment: Are Men and Women Different? examines data to compare men's and women's attitudes about spanking. This exercise is designed to give you practice in interpreting a contingency table and a test of statistical significance.
- Exercise 9.3 What Is Going On in the Schools? considers a study to see if African American and Hispanic students are more likely to be suspended than other students. This exercise is designed to give you practice in interpreting tests of statistical significance.
- Exercise 9.4 How Groups Work Together presents data comparing perceived characteristics of collaborations formed around women's issues and environmental issues. This exercise is designed to give you practice in interpreting tables that report the results of *t*-tests.

EXERCISE 9.1   Reviewing Polling Data

Scenario

You are a member of a community action group that focuses on various issues that affect community life. As you are perusing the Web for data on incidents of racial discrimination, you find a Washington Post–ABC poll of 1,079 randomly selected Americans. Two hundred and four of the poll respondents were African Americans.

Section A: Getting Started

1. The following data were reported in answer to the question "How big a problem is racism in our society today?"

   Of all respondents: 26% a "big problem," 22% a "small problem"

   Of African American respondents: 44% a "big problem," 11% a "small problem"

   Of White respondents: 22% a "big problem," 23% a "small problem"

   a. Use the reported statistics, for example, 26%, to estimate p and $1-p$. Compute the sampling error for the entire sample, the African American sample, and the White sample. (Assume that the number of Whites is the same as the number of non–African Americans.)

   b. Use the sampling error and report the confidence interval for the percentage of African Americans and Whites who believed that racism was a big problem.

   c. What is the probability that your estimates in 1b are wrong? How did you arrive at this estimate?

2. Assume maximum variability (50-50 split)

   a. Compute the sampling errors for the entire sample and for the 204-member African American sample.

   b. As a general practice, would you analyze survey data using a 50-50 split or would you use the statistical analysis (question 1) to come up with more precise estimates? Justify your answer.

3. The survey also reported that among African American respondents 60 percent had personally felt that "a shopkeeper or sales clerk was trying to make" them feel unwelcome. You are curious if the same thing is true in your community. In trying to decide a value of p would you use 0.60, 0.50, or something else? Justify your answer.

4. The community action group considers replicating parts of a survey. What size sample is needed to have 2 percent, 5 percent, or 10 percent accuracy? (Note that accuracy is the same as sampling error.)

EXERCISE 9.2   Attitudes toward Corporal Punishment: Are Men and Women Different?

Scenario

A child care organization commissioned a random survey to identify attitudes toward child-raising. A topic of interest was the difference between the beliefs of men and women regarding discipline. The follow table reports data on men's and women's attitudes toward spanking.

Attitudes toward Spanking by Respondent Gender

Favor Spanking	Male	Female
Strongly	115	107
Somewhat	212	221
Oppose	73	109
Strongly oppose	18	42

Chi-square = 13.1, degrees of freedom = 3, significance = 0.004.

Section A: Getting Started

1. State the alternative hypothesis and the null hypothesis that could be tested with the data in this table.
2. Identify the independent and the dependent variables.
3. Calculate percentages and include in them in a table. Write a sentence to describe the relationship shown in the table. Do the data in the table support or contradict your hypothesis? Explain.
4. Based on the chi-square evidence what would you do, that is, would you reject the null hypothesis?

Section B: Small Group Exercise

1. What are the implications of the findings presented in exercise 9.2? Do you consider the table an interesting observation, a question for further study, or something else?
2. A finding of statistical significance can be persuasive. What other evidence should the child care organization present so that parents and other stakeholders are able to evaluate the findings?

EXERCISE 9.3   What Is Going On in the Schools?

Scenario

A community action group has heard complaints that African American and Hispanic students are more likely to be suspended (either in-school or out-of-school) than other students. The superintendent of schools offers to review the files of students in grades 9–12. The school system has 39,000 students in grades 9–12: 58 percent African American, 12 percent Hispanic, and 25 percent White.

Section A: Getting Started

1. What target population would you recommend the superintendent use? Why did you recommend this population? (Note that target population refers to the specific population that the data will represent.)
2. State the alternative hypothesis and the null hypothesis the superintendent should test.
3. If the superintendent tests the hypothesis and makes a Type I error, explain what has happened.

4. If the superintendent tests the hypothesis and makes a Type II error, explain what has happened.
5. Should the superintendent be more concerned about a Type I error or a Type II error? Justify your answer.
6. The superintendent originally set $\alpha = 0.05$. How can she further decrease the probability of a Type I error? How can she further decrease the probability of a Type II error?

Section B: Small Group Exercise

1. Decide on a target population and suggest a sample size for the superintendent's study. Justify your recommendation.
2. Discuss what actions the superintendent might take if the null hypothesis is rejected.
3. List arguments for the position
   a. Committing a Type I error is the more serious concern.
   b. Committing a Type II error is the more serious concern.
4. Based on what you have observed in Exercises 9.2 and 9.3 draft a memo "What you want to know about statistical significance: A guide for citizens."

EXERCISE 9.4   How Groups Work Together

The following table was created as part of a study of collaborations formed around women's issues and environmental issues. Members answered a series of questions to see if the coalitions were different. Each question was answered along a scale ranging from 1 = Not at all true to 7 = To a great extent true.

Indicator	Women's Issues Mean(s)	Environmental Issues Mean(s)	t-value
• I understand my organization's roles and responsibilities	6.6 (.7)	5.9 (1.2)	1.7
• Partners agree about goals	6.6 (.5)	5.3 (0.8)	4.6***
• Meetings accomplish what is necessary for the collaboration to function	6.9 (.3)	6.1 (0.8)	3.1**
• Tasks are well coordinated	6.4 (.8)	4.8 (1.5)	3.1**
• Collaboration hinders my organization from achieving its mission	1.6 (1.2)	2.1 (1.5)	.9
• Collaboration affects my organization's independence	2.1 (1.5)	2.6 (1.6)	.8
• I feel pulled between meeting my organization's expectations and the collaboration's expectations	2.5 (1.9)	4.0 (1.5)	2.0**
• Partner organizations have combined and used resources so that all partners benefit	6.8 (.4)	5.5 (1.0)	4.0***
• My organization shares information that will strengthen the partners' organization	6.7 (.5)	6.1 (0.7)	2.3*

Indicator	Women's Issues Mean(s)	Environmental Issues Mean(s)	t-value
• Partner organizations appreciate and respect what my organization contributes	6.7 (.5)	5.7 (1.)	3.0**
• My organization achieves its goals better through collaboration	6.7 (.6)	5.9 (1.4)	1.7
• My organization believes it is better to stay and work with partners rather than leave	6.9 (0.3)	6.0 (1.0)	2.9**
• Representatives of partner organizations are trustworthy	6.9 (0.3)	5.4 (1.0)	4.8***
• My organization can count on each partner to meet its obligations	6.5 (0.5)	4.8 (0.8)	6.0***

Two-tailed t-test: *$p < 0.05$, **$p < 0.01$, ***$p < 0.001$.

Section A: Getting Started

1. In plain English explain what information the table contains.
2. In plain English interpret the statistical information for the last line ("My organization can count on each partner to meet its obligations.)
3. How do collaborations focused on women's issues differ from collaborations focused on environmental issues? What criteria did you use to make your choices?

Analyzing Interviews and Open-Ended Questions

You have designed your study, collected lots of information (maybe even got to meet some really interesting people), and analyzed your quantitative data. Now what? If your survey had any open-ended questions or if you conducted interviews or focus groups you may have lots more data to analyze. But how? We're glad you asked! Organizing and analyzing interviews and answers to open-ended questions require you to work with words instead of numbers. You need to decipher meanings from the accumulation of words that you, and perhaps others, gathered and recorded from interviews, focus groups, or surveys and link them to your research question.

At the outset, qualitative analysis may seem daunting. Some interviews can generate pages and pages of text and it may be a little intimidating to have to read, understand, and decipher the meaning of all of those words! But not to fear, while qualitative analysis takes time and effort, it is doable. This chapter will discuss how to analyze qualitative data in general. While most of the discussion will focus on analyzing data from interviews and focus groups, the techniques used are also applicable to analyzing open-ended questions from a survey or questionnaire.

Before we continue, it is worth giving a few words of wisdom. In Chapter 7 we talked about collecting only necessary data. It is worth repeating that admonition here. Many people add qualitative questions to their surveys because "it might be interesting" or decide to use qualitative methods because they "do not like statistics." Well, here is the trade-off. While statistical analysis can be intimidating, you can complete it relatively quickly—especially with the assistance of computer software packages. On the other hand, qualitative analysis can rarely be completed quickly. As a matter of fact, it is terribly time consuming and requires quite a bit of judgment and a degree of comfort with ambiguity. Rarely will you be able to conduct high quality qualitative analysis in a few hours' time, so keep that in mind as you construct questionnaires and ask interview or focus group questions.

Do not simply add a question because it might be interesting. Be sure to include only questions that will inform your research question and those that you will actually analyze. Now we're ready to talk about *how* to analyze the data you have collected.

Qualitative data often provide a deeper understanding of the phenomenon under study. Research participants have an opportunity to tell their story in their own words, on their own terms. These findings tell us not only the outcome but also often how it came to be, who was involved, and what the consequences of actions may have been. For example, a quantitative survey might tell us who started a program, how long it took them to complete it, and what they did afterwards while qualitative data will enrich our understanding of that process. Findings from qualitative data often shed light on the black box of quantitative evaluations. For example, suppose that a quantitative evaluation of an adult education program found that participants in one program outperformed their peers by 90 days in time taken to complete the program and get a GED. But this finding may be only part of the story. A qualitative analysis may find that a particular staff member inspired participants to do their best. All of the participants commented that she "made me believe I could do it." A quantitative survey may never reveal the depth of this person's impact.

Qualitative research has more credibility if data come from diverse data sources, such as interviews, focus groups, and observations. Your findings will be stronger and more convincing if they are supported by data from multiple sources. Presumably you will have conducted several interviews or focus groups. Recall from Chapter 7 that the recommended practice is to stop conducting interviews when you reach saturation, that is when you stop hearing new information.

PREPARING DATA FOR ANALYSIS

Before you begin your analysis give each source a unique identifier. Each questionnaire, respondent, focus group, or site should receive an ID number. The list linking ID numbers and respondents should be kept separately from the materials you will work with. This will help maintain confidentiality and decrease a possible source of bias in your data analysis.

Qualitative data are generally collected as notes, recordings, or transcripts. Interviewers may take written notes while interviewing a participant. These notes help interviewers remember important ideas during the course of the interview and after. The notes may help the interviewer capture major ideas, striking words, and issues that she may come back to during the interview. While note taking may be a wonderful supplement to recordings and transcripts, it is rarely the only method used. Even a skilled interviewer will find it difficult to take copious notes and conduct an effective interview. Most often either the interview or the quality of the notes may suffer. Still, notes are very informative and supplement recordings and transcripts.

Audio and video recordings provide reliable documentation of the participants' responses. Recordings provide an opportunity for the interviewer to focus on the conversation rather than capturing words. A digital or analog device can be used unobtrusively to record the interview. You must, however, obtain permission to record the interview. This permission should be part of the recorded interview as well as included on the consent forms discussed in Chapter 3. Most of the time, the recorder's presence recedes into the background and is forgotten during the course of the interview. If you have to turn over a tape, replace batteries, or otherwise tend to the recorder, do so with little fanfare. Most people understand the challenges of technology but you can minimize the interruption by having a backup device, tapes, batteries, or other power supply. While tape recorders might seem outdated, experienced interviewers continue to rely on them even if it is just as a backup to newer digital recorders.

Transcripts are developed from recordings. You can work directly from the recordings, but you may find it more time consuming. For example, you may waste a lot of time listening to the recording to locate a quote that you want to use. You may find that reading the written record of the interview will give you different insights into what was being said. You may complete the transcription yourself or hire a professional transcriber. In either case you need to decide if the transcript should be done verbatim or in formal English. Verbatim transcripts will have all of the "uhs" or "ums" of the participant. A verbatim transcript also provides non-word, verbal cues such as a shrug of the shoulder or rolling of eyes, which can add to the analysis. Including these nonwords in the transcript may cue the researcher to areas of ambivalence in the participant's response.

As you begin analyzing paper transcripts, recordings, and electronic files you may compromise your records. Transcript pages may get separated or electronic text may be accidentally deleted. Therefore, prior to beginning your analysis, make copies of all paper and electronic data and store them in a secure place. You should have one copy to work from and one copy as a backup.

CONDUCTING THE ANALYSIS

Reviewing the Data—Get to Know Your Data: Read, Reread

Qualitative analysis depends on your being intimately familiar with your data. Take the time to read all of the transcripts, listen to all of the recordings, and read your notes. As you do your first read, do not make notes. Simply read the information to *hear* the story that is being told by your participants. Review all of the data from each source before you begin your analysis. This is particularly important if more than one researcher has collected the data. While you may be familiar with the interviews or focus groups you conducted, you may not be as familiar with those that your colleagues or research partners conducted.

If your data were collected from open-ended questions, compile all the qualitative responses by question. For instance if you surveyed 100 school principals about parental involvement in schools and if question 25 was "Tell us about the strategies you use to encourage parental participation at your school," keep all 100 responses to this question together. It often helps to create separate files for the responses to open-ended questions and focus on one data set at a time. Once you have all of the data compiled, review all of the responses. As you become familiar with your qualitative data (whether obtained from surveys, interviews, or focus groups), themes will begin to emerge.

Coding the Data

Coding refers to the identification and classification of themes, concepts, behaviors, interactions, incidents, terminology, or phrases in the data. The process of creating and assigning codes to the responses will force you to think and rethink what the data mean. You may start with themes in mind and alter them as you code. You may notice unexpected linkages between behaviors. You may deepen your understanding of familiar concepts. Codes may be noted in the margins of the transcripts—another reason transcripts are preferred. Most often, the code is noted next to the text that represents the category.

You can begin by developing a preset list of categories that you are looking for in the data. Because you have listened to the recordings and read the transcripts you are somewhat familiar with the data. You may want to jot down a list of topics or themes that you encountered in the data. This process develops a preset list. You can assign the topics or themes with a short name or word; the name or word is called the *code*. When you come to a portion of the text that contains a relevant theme, you may enter the code in the margins. Invariably there will be other codes that emerge as you reread the transcripts or listen again to the recordings. You want to maintain flexibility and allow for additional codes as you engage with the data. These codes may appear as you become more familiar with the data and more sensitive to what respondents were trying to communicate.

You need to create a separate record of your codes; the record may be called a *codebook* or *data dictionary*. Normally you will note the code word(s) you used and a short definition or explanation. The definitions will help you remember what you meant by a particular term. You may find that as you read more and think more about the codes and categories you need to separate a category or condense a category. Each interview or focus group transcript should be individually coded. Coding is not a onetime process; nor is there always only one code per portion of text. Coding requires significant time and effort: it is a flexible, recursive process, that is, as you work with the responses you may reconsider and adjust your coding scheme.

Another consideration is identifying manifest and latent content. *Manifest content* is observable and is on the surface of the data. Manifest content lends itself to word counts and easily identifiable phenomena that require no interpretation. *Latent content* is the researchers' interpretation of the meaning of the data. For instance, the following portion of a transcript is from two school social workers discussing parental involvement in schools:

CYNTHIA W: . . . I don't think the teachers understand. . . . Kids that have to get up and get themselves on the bus at 6 A.M. themselves because their mama has to go to work. . . . People tend to think that everybody's life is like your life. My husband is the same way . . . he can't relate to what it's like now. It's hard. I couldn't do it. People just don't understand how hard these parents have it!

CAROLINE M: And there's some [parents] that just need to be pushed to do what they are supposed to do! They are just not self-starters; they need somebody to get them on track. Maybe they don't have that person there to do it for them, so they don't. I would love to see more parent involvement, and more resources to be able to offer to parents. To assist them, I feel that we do have a lot of parent involvement, yet, there's a lot of situations where you work with the same families year after year. And you're kind of like back on square one . . . and it's either they are not willing to change or not able to change. I just don't understand! So sometimes you feel like you're up against a brick wall, and you can't understand why they won't [complete procedures to obtain social service assistance]! So I would like to see a lot more parental involvement.

Reading the information above, several words appear repeatedly. For instance, both speakers say the word *understand*. The word is used four times in this brief encounter. It represents manifest content and may coded as *"understand."* Latent content can also be identified. A researcher might identify "frustration" and "parental struggles" as latent content. Because latent content is interpreted by the researcher, it is more subjective than manifest content. If more than one researcher is coding transcripts, you should use inter-rater reliability to confirm the coding, especially the coding of latent content.

Analyzing the Data: Second Level Coding

Now that you have created categories for the data, think about comparisons and relationships between categories. How do the categories link, if at all? Do some categories complement or contradict others? Do some large categories have lots of references in the text and others are found one or two times? Are some categories more essential to understanding an issue than others? Categories may be temporal, embedded, or unrelated. That is, some phenomenon in the data may happen before another (temporal), other events might be part of another (embedded), and some categories may stand alone. The following examples describe each category.

In a study about survivors of Hurricane Katrina several codes were created, including resilience, depression, gastro-intestinal complications, stress, family support networks, being labeled as "refugees," and insomnia.

- *Temporal.* The respondents all mentioned reconnecting with family support networks and church as a precursor to their ability to begin to rebuild (resilience).
- *Embedded depression.* Gastro-intestinal complications and insomnia were found to be interconnected. The initial codes were ultimately condensed to become mental and physical health.
- *No relationship.* Being called refugees was not related to the other codes.

Finding these relationships between categories is called *second level coding*. You can use low- or high-tech methods to create diagrams similar to that in Figure 10.1, to describe the relationships. Simply draw the relationships on a piece of paper or use a computer program (word processing, desktop publishing, and presentation software packages all have this capability) to help you create a diagram so that you can see the relationships.

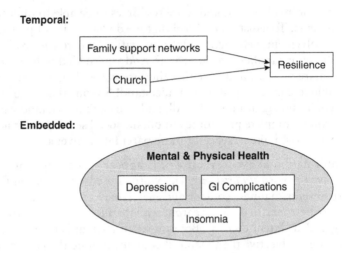

Temporal:

Embedded:

FIGURE 10.1
Diagrams to Illustrate Relationships Between Categories

Interpreting the Data

Although you have now identified the relationships between categories and ideas generated by your analysis, you still need to make sense of it all. If you are doing exploratory research your final step may be to describe each component or category. You may be tempted to list the findings without providing any synthesis and discussion of the implications of the categories and the relationships. Interpreting the data requires attaching meaning and significance to the codes and categories: to tell the story as you now understand it.

Qualitative findings are used to generate recommendations and theories. A first step is to think about what you have learned. What lessons have emerged from your analysis? What did you learn that was new or different? What would you suggest to others exploring the same programs or services? Might your findings have relevance in other similar settings? Does what you have found conflict with or support existing literature on this topic? Are there gaps in your findings or areas that remain unclear? Do questions for additional research emerge from the current findings?

Your interpretation of the data is a thought process. You are communicating to your audience what you now know and how it can be built upon to develop new theories, programs, or systems. The example of the evaluation of the adult education program revealed the importance of a particular staff member's actions and her impact on program participants. A possible interpretation of that finding is that it is important to have staff provide affirmations and positive reinforcement to program participants.

Using Quotes to Support Your Points

A beginning qualitative researcher may be frustrated with the lack of explicit structure for interpreting qualitative data. There are very few "hard and fast" rules for interpreting data. If you use quotes and check for bias you will have made a good start in rigorous interpretation and producing work that withstands critical examination.

In qualitative data analysis, quotes are often used to support your interpretations and illustrate your observations. Readers may find a direct quote from the transcript quite compelling. You should be careful in choosing quotes. The quotes should be directly relevant to the point you are trying to make and they should add to the discussion. Avoid using long quotes that go off on tangents. They will only distract the audience or dilute your point. You should be able to justify why a specific quote was chosen and what it adds to the discussion.

Be careful to use a balance of quotes. If everyone did not agree with a specific point, be sure to add quotes not only from those who agreed but also from those who disagreed. Don't use only quotes that support your assertions. Make it clear that while this quote represents your conclusion, it is not representative of all of the points of view.

When using quotes be careful not to take them out of context. When using direct quotes make sure that you are not editing or cutting off quotes and leaving out important contextual cues that would clarify the statement. Sometimes you will need to include the interviewer's question in order to make sure that the full meaning of the quote is captured.

Remember that using a pseudonym may not be sufficient to protect subjects' confidentiality. Some information provided by the participant may be so specific that the participant can be identified by the quote. A consultant doing a report from a strategic plan for a small organization once began a quote saying, "A female worker on the child welfare project does not feel that management appreciates the contribution of women." Only one woman was employed on the child welfare program! While the worker's name was not provided, a reader familiar with the organization could have identified her. The writer could have put her at risk for retaliation.

Biases and Credibility

Because qualitative research is subjective, there are many opportunities for it to be done poorly. As with any research endeavor, bias can influence your findings or your interpretation of those findings. However, several strategies may reduce the impact of bias and increase the credibility of your qualitative research.

- *Take stock of and understand your biases.* Be aware of your own preconceived notions, biases, and even prejudices before you begin the data analysis. Write them down and refer to them often. Make sure you are not putting your values and beliefs on your participants' words and how you interpret them.
- *Remember your question.* To make sure you do not go off on a tangent, make sure that your analysis hinges on your question. Because of the enormous amount of data generated by qualitative research you may lose sight of the purpose of your study and explore other interesting but not terribly relevant ideas.
- *Exhaust the data.* While you want to remember your question and avoid tangents, this can be done only by coding and working with all your data.

You have to code all your data to know what is relevant and what is irrelevant to your question. Keep in mind that information that contradicts an expected finding is not irrelevant. It must be included and discussed. Remember, these contradictions are often the beginnings of new ways of looking at old problems.

- *When feasible, work with a team.* As with quantitative research, inter-rater reliability is useful in qualitative research. Have another colleague on the research team independently code the data. This separate review of the data will help solidify the codes on which you agree and will require you to clarify codes on which you disagree. Including colleagues will help in developing theories and determining meaning from the data.

Use *member checks*, which are similar to the team approach. You can arrange to have "members" or the study participants review your findings and interpretations. This strategy improves the validity, accuracy, and credibility of data in qualitative research. While many researchers conduct member checks as they write their final report, there is no universally accepted point in the research process for this activity. Member checks can be done during the interview process, to confirm the data's accuracy. It can also be done during analysis to confirm the validity of the findings and/or after the report is written to provide a point of closure for the participants and researchers. There is an important caveat to conducting member checks. You should remind respondents that the findings are the aggregate of all the responses and therefore may not exactly reflect their comments or points of view. However, if a respondent feels strongly that an important point has been omitted—or conversely, overemphasized—you may follow up and verify the concern. This situation makes a case for conducting member checks *before* writing the final report.

Computer-Assisted Data Analysis

With technology as prevalent as it is, computer software may be used to help with data analysis. While the steps are the same, software programs can often further classify and organize data. There are several qualitative research software packages, such as HyperRESEARCH, QSR Nvivo, and Atlas.ti. Each software package has features that will help create order to the data, such as list the codes, code the text of transcripts, count code frequencies, and write memos for future reference. Currently available software may cost more than an organization is prepared to spend. Not to worry. Word processing packages can accomplish the same tasks. You can use the search function, change text colors, and insert comments as you track changes. By using word processing software, you can avoid purchasing special software, especially if the quantity of qualitative research does not justify the cost.

CONCLUDING OBSERVATIONS

This chapter discusses how to analyze data obtained using qualitative methods. While analyzing these data may be a bit overwhelming at the outset, their potential for providing additional detail, assessing operational validity, and increasing creditability is worthwhile. Qualitative research done well can provide a glimpse into the whys and wherefores of larger trends, provide explanations that are not readily apparent, and explore uncharted territory. Qualitative data analysis must be done systematically and carefully. The process of analysis includes preparing the data; reading responses to become familiar with the data; coding; classifying themes, concepts, behaviors, and phrases; and finally exploring the interactions between codes to interpret meaning.

Two major points are essential for you as a researcher or manager to remember: First, understand that this is a process and second, watch out for biases. Data analysis is an iterative process that does not lend itself to simplistic conclusions and determinations. It cannot be done quickly, its findings are not unambiguous, and it requires you to engage with the data. To truly reap the benefits of qualitative data, you must continually ask "what does this (what all these people have said) actually mean?" and seek understanding from all those words. You will have to think about and reflect on the emerging themes and often adapt and change your previous assumptions as you interact with the data. The theories that emerge as you collect and analyze the data should be challenged and refined throughout the process.

Also remember that while there are lots of electronic tools (e.g., recorders, computer-assisted software) to help in the data analysis process you are the main interpretive tool. Therefore, you have to be careful not to let your "stuff and baggage" taint the lens through which you interpret the data.

RECOMMENDED RESOURCES

Gibbs, G., *Analyzing qualitative data. The Sage Qualitative Research Kit* (Thousand Oaks: Sage Publications, Inc., 2007).

Rapley, T., *Doing conversation, discourse and document analysis. The Sage Qualitative Research Kit* (Thousand Oaks: Sage Publications, Inc., 2007).

Patton, M. Q., *Qualitative Research and Evaluation Methods* (Thousand Oaks: Sage Publications, Inc., 2002).

Saldana, J., *The Coding Manual for Qualitative Researchers* (Thousand Oaks: Sage Publications, Inc., 2009).

CHAPTER 10   EXERCISES

This chapter contains two exercises.

- Exercise 10.1: Nonprofit Lobbying asks you to analyze a transcript to learn why nonprofits do not lobby.
- Exercise 10.2: Learning from Others—Making a Public Apology instructs you how to craft a public apology. This exercise asks you to locate and analyze transcripts to help you design an effective risk management strategy.

EXERCISE 10.1   Nonprofit Lobbying

Section A: Getting Started

1. Recently interviews were conducted as part of a study about why nonprofit organizations do not lobby their legislators for programmatic and fiscal policy. Take a moment to develop a list of preset codes you will use to code the interviews. Write a definition or brief phrase explaining each code. Do this by considering the varied reasons nonprofit organizations might refrain from lobbying.
2. Read the following excerpt from an interview transcript. Using the margins, code the data using your list of preset codes. If additional codes emerge, note

them on your list of codes and provide a definition for each new code. Make five copies of your code list.

INTERVIEWER: I am going to record our conversation so that I can actually pay attention to what you're saying and then can actually listen rather than scribble notes. Is that okay with you?

RESPONDENT: Okay. I give my permission to be audio-taped.

INTERVIEWER: Thank you. And if you could just state your full name and your title for the recording, and then we can go ahead and get started.

RESPONDENT: Okay. I'm Stephanie Martine. I'm executive director of Family Services Nonprofit in South Raleigh, North Carolina.

INTERVIEWER: Okay, thank you. What we'd like to do today is to just have a brief conversation with you about the status of nonprofit lobbying and advocacy. We wanted to start off the conversation by asking, when I say "nonprofit lobbying and advocacy," what comes to mind for you?

RESPONDENT: I guess it's making sure that the laws are supportive of nonprofit agencies in the United States.

INTERVIEWER: And is there a difference between lobbying and advocacy as you see it?

RESPONDENT: [Pause] Lobbying deals more with the laws. Advocacy would be more dealing with rights.

INTERVIEWER: Okay. And does your organization, as you think about what you do, do you feel like you engage in lobbying and advocacy?

RESPONDENT: No.

INTERVIEWER: Okay. And what keeps you from doing that?

RESPONDENT: We're a very small agency. We have all part-time staff.

INTERVIEWER: Okay, what's the size of your organization?

RESPONDENT: We have four part-time staff.

INTERVIEWER: And your position is also part-time?

RESPONDENT: Yes.

INTERVIEWER: And so it's really a time issue for you?

RESPONDENT: Yeah, any extra time I have I'm writing grant proposals to keep going. [Laughter]. I feel like I just run to keep my head above water. And I think our board has a fear of being misquoted, misrepresented, and so there's an attitude of "stay under the wire" there. [Laughing]

INTERVIEWER: So fear is a major issue?

RESPONDENT: And then there's also a fear that if we publicize what we do too much we don't want to create waiting lists. We don't want to be in a position where we have more demand for our services than we have services available.

INTERVIEWER: What would make it easier for you?

RESPONDENT: Hmmm . . . to make things as quick and easy as possible for people to respond to, adding your name to a petition is easier than sitting down and writing your own letter and sending it. Even when somebody sends you a prewritten letter sometimes that's even hard. Whereas if I can hit "reply" and say "Yes, me, too." You know?

INTERVIEWER: Tell me more about that.

RESPONDENT: I don't know. I mean, sometimes I walk away feeling guilty because I wanted to, and I felt that way, too. But I just didn't have one more second to attend to it. I would think it's time.

INTERVIEWER: Any other reasons?

RESPONDENT: Time, and of course money. Not having the money to hire a lobbyist, or garner all the resources that it would take. But also fear, just kind of being afraid of the whole idea. I think that that has merit. And I do think there's some misinformation, miseducation of what you can or can't do in regards to lobbying or speaking out for things. People just don't understand, or they've never been given specifically, "Well, here's how you can do it." And so maybe that's the fear. I mean I guess I have the fundamental belief that our politicians want to hear from us. They want to know what's going on, and they're relying on the public to keep them informed.

INTERVIEWER: In a perfect world, beyond time and money, what would help you to lobby?

RESPONDENT: Knowing the rules. I just am not sure what we can do and what we can't. It gets so frustrating. You go to workshops and get conflicting information so that's just a waste of time too.

INTERVIEWER: Anything else?

RESPONDENT: To be honest, sometimes I'm a little intimidated by going to the legislature, especially in Washington, D.C. All those people in their high powered positions. Their slick suits and shiny shoes. I'm just a part-time director. We can't grease their palms like those lobbyist do. [Deep sigh] We're not in the same league.

Section B: Class Inter-rater Reliability

In groups of three to five provide copies of your code list and code definitions to your group members.

1. Compare the code lists. Come to group consensus on the list of codes and their definitions.
2. Write or draw out the interconnections of the themes. Be sure to
   a. condense codes to develop larger themes;
   b. identify any temporal codes;
   c. identify any embedded codes.

3. Using the developed themes, identify at least one theory about nonprofit lobbying.
4. Identify at least two quotes from the list to support your theory.

EXERCISE 10.2 Learning from Others—Making a Public Apology

Scenario

The executive director and the board of a small nonprofit agency have been working on risk management strategies, including the possible risks involved in giving public statements. You have been asked to craft a public apology statement. You decide to start by studying the structure of a public apology statement. Think about well-known public apologies in the news and conduct an Internet search for the speech transcripts or videos. (Some notorious public apologies include those given by Tiger Woods, John Imus, Mel Gibson, Bill Clinton, Chris Brown, Michael Vick, Kevin Rudd, Carly Fiorina, and Akio Toyoda.)

Section A: Getting Started

Choose at least four videos of public apologies. You can search the term *public apology video* in a video Web site or general Web site. You will get the most recent apologies. Choose more than those from the latest news scandal. Make sure the videos you choose have enough content to provide several possible codes. Try to find videos that are at least 3 minutes 30 seconds long. Create a verbatim written transcript of each video. Be sure to note specific gestures made by the speaker. For instance, if the speaker points to the camera, folds his or her arms, or rubs his or her face, make that notation in parentheses in the transcript.

1. Develop a list of expected codes but be sure to remain open to new codes that emerge from the transcripts.
2. Code the transcripts based on latent rather than manifest content.
3. Detail the interactions between the codes and/or categories.
4. Identify at least two quotes to serve as examples of each category.

Section B: Small Group Exercise

Working together in groups of three or four build on what each of you has learned about public apologies.

1. Compare your latent content codes.
2. Negotiate among yourselves on the relevant categories.
3. Develop a theory about the structure of a public apology statement.
4. Write a template for a public apology.

Programs and Communities

The chapters in Section IV cover a variety of topics and thus are not as closely tied together as those in earlier sections. Organizations routinely compile performance data and conduct surveys. Program evaluations and needs assessments are conducted less frequently; they may be conducted by a specialized team within the organization or by contractors. You may find yourself contracting for a study, participating on a team, or assessing the findings. Geographic information systems—GIS—is an emerging research strategy that can be used for a variety of purposes. We included it here because it can be applied to either program evaluation or needs assessment.

Research to Evaluate Programs

How often have you heard politicians or new agency heads say, "We want programs that work!" Yet, deciding that a program works is more complicated than you or they may think. To decide that a program works you need to identify and measure outcomes that indicate a program's effectiveness and you need to examine the evidence that links a program's activities to its outcomes. Whatever your professional role, you will probably examine the operation and outcomes of programs. As an administrator you may want to establish that you have sufficient resources to implement a program, that its activities lead to desired results, and that it is efficient. You will want to provide program information to your board, donors, and the public. As a board member you will want to direct resources toward funding or advocating for effective programs. As a funder, whether you represent a foundation or a government agency, you may require evaluations to demonstrate that funded programs complied with donor or government requirements and achieved promised results. As a taxpayer you want to be aware of evidence that a program works or does not work and under what conditions it works.

To produce evidence of program effectiveness and efficiency you will want to conduct a program evaluation. Conducting a quality program evaluation is a tall order! Program evaluations may take many forms. Over time, program evaluation methods have expanded to include needs assessment, stakeholder opinions, implementation research, and cost-benefit analysis. Two common forms, directly related to program effectiveness, are process evaluations and outcome evaluations. Administrators use process evaluation to describe a program's operations. They may conduct a process evaluation to learn if a program was implemented as expected, if it served as many people as promised, and how much each unit of service cost. To conduct a process evaluation administrators measure and track the components included in a logic model, which we covered in Chapter 4. The data may come from a performance-monitoring system or surveys. Administrators conduct outcome evaluations to demonstrate that a program achieved the desired results or outcomes. They may, however, only identify and track outcomes, and ignore other factors that may have produced them. Such studies risk overlooking important variables that actually explain the outcomes. For example, the success of vocational training programs may partially depend on economic conditions: a high quality program may place few graduates during a recession, whereas a mediocre program may place most graduates during an economic boom. To produce more convincing evidence, investigators may design a study to rule out possible causes of outcomes other than program activities.

Process evaluations do not attempt to establish whether a program brought about the desired outcomes; consequently, they may not get the respect they deserve. However, process evaluations are valuable. You can use process evaluation information about program resources and activities to guide decisions whether to expand a program, where to offer it, and who should offer it. You may uncover problems such as too few participants, high drop-out rates, inadequate resources, or poor program implementation. Consequently, an outcome evaluation is more efficient if a process evaluation has already identified, defined, measured, and tracked program components. Without process information you risk basing your outcome evaluation on poorly defined outcomes or a partially implemented program. An outcome evaluation produced under these conditions will not provide useful information.

In this chapter we focus on evaluation approaches that produce evidence of program effectiveness. As you will learn, no one study provides definitive evidence of program effectiveness. First, you have the challenge of showing that the program and not something else caused the outcome. Second, each study involves specific subjects in a specific setting, so you cannot be sure that the findings from one study will apply to other subjects in other settings. Third, implementing a program widely may result in it gaining strong stakeholder support, in which case evidence of its ineffectiveness may fall on deaf ears. For example, evaluations of D.A.R.E. (Drug Abuse Resistance Education), a drug-prevention program for adolescents, consistently questioned its effectiveness, but to little avail.[1] The material in this chapter will enable you to design and critique an evaluation and to assess the evidence that any one study provides.

HOW TO FOCUS AN EVALUATION

To produce evidence that a program works you should begin with a program or logic model. In Chapter 4, Figure 2, we saw that the basic form of a logic model is

$$\text{Inputs} \rightarrow \text{Activities} \rightarrow \text{Outputs} \rightarrow \text{Outcomes}$$

Inputs are a program's resources. Activities are what the program does, that is, how it allocates and consumes the resources. Outputs are units of service or goods the activities produce. Outcomes are the results of the activities and outputs.

---

[1]For a discussion of D.A.R.E. and the influence of evaluation, see C. H. Weiss, E. Murphy-Graham, and S. Birkeland, "An alternative route to policy influence: How evaluations affect D.A.R.E.," *American Journal of Evaluation* (2005), 26:12–30. For a discussion of its stakeholder support, see E. Wysong, R. Aniskiewicz, and D. Wright, "Truth and Dare: Tracking drug education to graduation and as symbolic politics," *Social Problems* (2004), 41(3):448–472.

As you build a logic model you link the outputs to the desired outcomes; you link activities to the outputs they should produce and the resources (inputs) to the activities that consume them.

You can answer important questions with the information associated with each link: are resources used appropriately? What outputs do the activities produce? Do the outputs result in the desired outcomes? If the desired outcomes were not achieved, why weren't they? You can infer that a program failed because it had inadequate resources, it was not carried out as intended, or that the theory linking the activities to the outcome was flawed. Even if the intended outcomes occur, you need to develop evidence that that the program and not something else produced the outcomes.

DESIGNS FOR EVALUATION

Think of an action you have taken or wanted others to take with the hope that it would make a difference. Perhaps you designed a fund-raising campaign or developed a volunteer program. Were you successful? How do you know? You may have examined performance data such as the amount of money raised before and after the campaign or the change in the number and quality of volunteers after the program was implemented. These and other performance measures may provide useful—but inconclusive—evidence that your actions brought about the observed changes.

To test a claim that an action or program caused a specific outcome, you need evidence that

1. a statistical relationship exists between the action and the outcome;
2. the action, or intervention, occurred before the observed outcome;
3. other possible causes of the outcome have been ruled out.

To find statistical evidence of relationships between program activities and outcomes, you can analyze surveys and performance-monitoring data. To establish that an intervention occurred before an outcome, you can examine performance-monitoring data, and occasionally, surveys. Still, neither performance-monitoring systems nor surveys can eliminate other possible causes of an outcome. Ruling out other, alternative causes of an outcome is more difficult than simply demonstrating that the intervention occurred before the outcome. To help identify possible alternative causes, methodologists have identified categories of research problems and labeled them as threats to internal and external validity. Since the terminology can be confusing, let's discuss the concepts in each category and how they relate to program evaluation. *Internal validity* refers to evidence that a program, action, or intervention (and not something else) caused observed changes. *Threats to internal validity* refer to features of a study that may challenge (or threaten) the assumption that an intervention caused an outcome. *External validity* refers to the evidence that study findings apply to unstudied cases. *Threats to external validity* are features or the context of a study that limit the application of findings beyond a given study.

In the following sections we summarize the common threats to internal validity and present two randomized experimental designs: quasi-experimental designs and non-experimental designs. Randomized experiments do a better job than other research designs of demonstrating that a program caused a given outcome. We then discuss quasi-experimental designs, which have some, but not all of the characteristics of a randomized experiment, and non-experimental designs, which provide very limited evidence of a causal relationship. Next we briefly discuss external validity and meta-analysis. The best evidence of external validity is replicating a study in different settings with different populations. Investigators use *meta-analysis* to combine studies of an intervention to strengthen the evidence of program effectiveness. We conclude this chapter by briefly considering cost benefit and cost effectiveness analysis, which incorporate program costs as part of identifying programs that should be implemented.

INTERNAL VALIDITY

A threat to internal validity offers a plausible alternative reason why an outcome occurred. Categories of threats to internal validity are as follows:

History. Events other than the intervention that occurred at the same time and could have caused the outcome.

Maturation. Changes in program subjects that occur naturally over time.

Selection. Characteristics of program subjects that may have caused the outcome.

Statistical regression. Program subjects are picked because of an extreme value, such as a low performance score, and the outcome is due to an expected shift to typical performance.

Experimental mortality. Subjects drop out of a program before it is completed; if they are dropped from the analysis it may distort the outcome.

Testing. An initial measure or test that influences the subjects and causes the outcome.

Instrumentation. Changes in the outcome measure from one time to another or between groups that affect the outcome.

Design contamination. The context of the study that brought about the outcome: (1) participants may act differently because they are in a study; (2) participants may alter their behavior to influence the outcome; or (3) participants exposed to an intervention may interact with comparison group members, so that both groups are effectively exposed to the intervention.

To further explain how to identify and interpret a threat, let's consider history, maturation, selection, and design contamination in more detail. If you understand these threats you should be able to identify the other threats. Consider history as a possible threat if other events coincide with the intervention. For example, suppose that as a fund-raising campaign began, a major disaster occurred; the campaign resulted in record donations. True, the campaign would be statistically related to the amount raised and it occurred before the money was collected. Yet, you cannot eliminate the disaster as a cause for how much people donated and to whom they donated. Similarly, policies to address a serious social problem seldom happen in isolation. A region experiencing severe drought may quickly enact policies to reduce water use: increased fines, higher water fees, and incentives for low-flow toilets and shower fixtures. If water usage drops it would not be clear which strategy or combination of strategies brought about the change.

Consider maturation a possible threat if an outcome may have resulted from natural changes that occur over time. You should suspect maturation as a threat to the internal validity of before and after studies. For example, we expect reading scores of elementary school children to improve as they progress through a school year. Similarly, you may find worker productivity varies near the end of the day or as the weekend approaches. The changes may be due to the natural changes in developmental skills, alertness or mood, respectively. To conclude that an intervention caused an observed change, you need to demonstrate that the change is greater or less than one would expect "as time goes on."

Consider selection as a possible threat if the characteristics of the subjects may explain the outcome. For example, a training program may recruit recently laid-off workers. If these trainees return to their old jobs or similar jobs, the program may take credit for the placement, but you may question if the trainees' experience and not the training led to their getting hired. At one time hormone replacement therapy (HRT) was widely prescribed because women receiving HRT had fewer heart attacks. Critics argued that the women who took HRT had higher incomes and better medical care, which contributed to their better health. Subsequent research, which did not have a selection bias, did not support the link between HRT and heart health.

Design contamination covers other factors that weaken evidence of causality. Participants in a study can be expected to be self-conscious and act differently than they normally do. People may answer questions to present themselves in a certain way; they may take pains to give answers that suggest they are tolerant, optimistic, or capable. Participants may try to achieve an outcome in their best interests. This may be particularly true if participants believe that a study may lead to a change in how they do their job or whom they work with.

Subjects exposed to an intervention may talk to those not exposed to the intervention. Consequently, both groups may experience the same or similar intervention. Consider an experimental weight loss program. Potential participants are randomly selected to participate in the program. If the program participants lose weight they may share the program's information and strategies with others. As a result the program may seem unsuccessful because both participants and nonparticipants lost weight. In this case you may assume the weight-loss program had no effect when in fact it did.

As you design a study or review research keep the threats to internal validity in mind. You can use the threats as a checklist to help you rule out or consider alternative explanations of why an outcome occurred. You should not fret over correctly naming a threat; you only need to identify plausible alternative hypotheses. You may not be able to eliminate all threats. You should aim to identify threats that are so grave as to make the findings worthless. You may conclude that other threats weaken the evidence of a program's impact and that you need further evidence. You may consider some other threats, while real, as inconsequential.

EXPERIMENTAL DESIGNS

Experimental designs provide the strongest evidence that an intervention caused an outcome. The key feature of an experimental design is random assignment. You randomly assign subjects to an experimental or a control group. Then you expose the experimental group, but not the control group, to the intervention. At the end of the study you determine if the two groups have different outcomes. You should make every effort to keep the experimental and control groups similar at the beginning, during, and at the end of the study. The only difference between the two groups is that you have exposed only the experimental group to the intervention. Random assignment eliminates selection and statistical regression as threats to internal validity. Each subject has the same probability of being in either the experimental or the control group. Because of random assignment you can assume that both groups were similar at the beginning. Nevertheless, you need to know more about the details of the study to ascertain that the initial similarity continued. For example, the experimental group, and not the control group, may have experienced events other than the intervention that caused the change (history). Subjects may have dropped out (experimental mortality). Design contamination may have eliminated the differences between the groups or affected the groups' behavior.

The two basic experimental designs are the classical experimental design and the randomized posttest-only design. Being familiar with these designs will help you understand and interpret other designs that build on them.

The Classical Experimental Design

The classical experimental design is the prototype of a randomized experiment. Its key steps are to

1. randomly assign subjects to either an *experimental* or a *control group*. With random assignment each subject has the same chance as any other subject of being in either the experimental group or the control group;
2. administer a pretest, measuring the outcome, to both groups;
3. expose only the experimental group to the intervention. Other than the intervention both the experimental and control groups should experience the same conditions;
4. administer a posttest, measuring the outcome, to both groups;
5. determine the amount of change in the outcome between the pretest and posttest values for each group. The difference between the two groups can then be attributed to the intervention.

Recall that evidence of causality requires that the intervention and outcome are statistically related, that the intervention occurs before the outcome, and that alternative hypotheses have been eliminated. Step 5 establishes a statistical relationship. Steps 2, 3, and 4 require that the intervention occurred before the outcome. Steps 1 and 3 seek to eliminate alternate hypotheses. The hardest threat to control may be design contamination. You may reduce design contamination if you keep the experimental and control groups from interacting. If an intervention takes place over time participants may drop out, introducing experimental mortality as a threat.

To identify a study's components and the sequence of events you may find the following standard notation helpful. The notation tracks the number of groups, how members were assigned to each group when observations were taken, when an intervention was introduced, and who was exposed to the intervention. The standard notation uses R to indicate that subjects were randomly assigned to an experimental or control group. O represents the value of a measured outcome and X indicates the intervention. O and X are organized along a line that represents the time sequence of when the measures were taken and the intervention was introduced. Subscripts for O indicate the time order; that is, a subscript 1 indicates the first time the outcome variable is measured, a subscript 2 indicates the second time it is measured and so on. The classical experimental design is represented symbolically as follows:

$$R \; O_1 \; X \; O_2 \; \text{(experimental group)}$$
$$R \; O_1 \quad O_2 \; \text{(control group)}$$

O_1 is the pretest and O_2 the posttest. The value of O for both the experimental and control group members may change between the pretest and posttest. All or some of the change may have been due to factors such as history, maturation, or testing. If both groups experience the same conditions and events, with the exception of the experimental group's exposure to the intervention, you may attribute the posttest difference between the experimental and control groups to the intervention. Because other than the intervention everything about the study conditions was the same for the two groups, you can eliminate alternative hypotheses as possible causes.

Consider an innovative 6-week program to improve nutrition. You randomly assign potential subjects to two groups. One group starts the program immediately and the other group is put on a waiting list. You survey both groups about their eating habits and their knowledge of nutrition. At the end of the 6 weeks you resurvey both groups. Random assignment assumes that both groups are similar and that the survey (test) and any publicity (history) impact both groups in the same way. If the survey and resurvey are given at the same time we can assume that changes over time (maturation) have had a similar effect on both groups.

Randomized Posttest Only Experimental Design

The randomized posttest design is less costly and easier to implement than the classical experimental design. You may assume that a pretest is unnecessary because random assignment should yield equivalent experimental and control groups. But you should not use this design if your study involves only a few subjects; you would then need pretest data to verify that the groups are equivalent. To determine if the intervention was effective the design calls for you to compare the outcomes of the two groups. The randomized posttest design is represented symbolically as

$$R \; X \; O_1 \; \text{(experimental group)}$$
$$R \quad O_1 \; \text{(control group)}$$

With a randomized posttest design you can demonstrate a statistical relationship between the intervention and the outcome. Since you will gather the outcome data after the intervention your study meets the time-order requirement. You can assume that random assignment controls for most threats to internal validity. The design eliminates the possibility that the pretest sensitizes subjects and acts as a contaminant. However, experimental mortality may be a threat if you cannot collect outcome data from all the randomly assigned subjects.

In comparing the classical experimental and the posttest only designs, you might wonder what value a pretest adds. It can corroborate the assumption that the experimental and control groups are similar. With a small number of subjects such corroboration is necessary. With a pretest you can better estimate how much of the difference between groups can be attributed to the intervention. In field experiments the equivalence of the groups may be lost as subjects drop out. The pretest data can help identify patterns among dropouts and infer the effect on the outcome data.

Randomized designs provide powerful evidence of a causal relationship between an intervention and an outcome. They work best if you can control the environment and assure that the experimental and control groups experience the same conditions other than the intervention. Such control is most possible with studies that involve relatively little time and occur in settings that you can monitor.

As an example of a simple intervention in an easily monitored setting, let's consider a campaign to improve the diet of expectant mothers. Suppose that study designers have created an information packet; it contains materials that picture ethnically diverse subjects, makes extensive use of illustrations, and has dietary suggestions on a wide range of food choices. To make sure the new packet is effective, investigators ask women waiting in obstetricians' waiting rooms to examine a packet with existing information (the control group) or the new packet (the experimental group). The investigators randomly distribute the packets to the women. After reviewing the materials, the women answer questions about what constitutes a healthy prenatal diet. The investigators compare the answers of the control group and experimental group. If women who receive the new packets are more knowledgeable, the sponsors may conclude that the packets are effective and that the campaign should go forward.

Even with such a simple experiment, experimental mortality or design contamination may occur. Experimental mortality may occur if some of the women don't complete the questionnaire. Design contamination may occur if the women interact. (Assembling the materials in packets may keep the subjects busy and less likely to talk and share information with each other.) If the investigators conducted a classical experiment they would have had to deal with the logistics of administering a pretest, preventing the women from interacting, and keeping them engaged until they completed the questionnaire.

Quasi-Experimental Designs

Often you cannot randomly assign people or groups to an experimental or control group. It may be unethical to randomly assign individuals to treatment and control groups; this, was an issue in the Tuskegee study discussed in Chapter 3. In addition you may not be able to maintain the design over time. As time goes by you may find it harder to argue that the two groups are similar except for the intervention. Participants may drop out. The design may be contaminated as the remaining participants seek additional programs or other sources of information. You can use alternative designs, called quasi-experimental designs, to obtain as much information as practical and rule out as many threats to internal validity as possible. These designs have some of the characteristics of an experiment but lack other characteristics. Most notably, these designs do not include random assignment; hence, R does not appear in the notation describing the designs. Two common, basic quasi-experimental designs are the comparison group design and the interrupted time series.

The *comparison group design* is similar to the randomized pretest-posttest design, only you do not randomly assign the members to the experimental and comparison groups. You may compare persons who have participated in a program with those who have not. You may compare sites that have implemented a program with ones that have not. The quality of this design depends on how similar the experimental and comparison persons or sites are. The design is represented as

$$O_1 \, X \, O_2 \text{ (experimental group)}$$
$$O_1 \quad O_2 \text{ (comparison group)}$$

For example, assume that several counties have a Good Diets, Healthy Babies campaign to reduce the rate of low-birth weight babies. To evaluate the program you identify and collect data from comparison counties. You might look for counties with similar demographics and the same number of medical services. However, your matches won't be perfect. You must be open to arguments that differences between the experimental and comparison counties are due to characteristics that you did not include in selecting the counties (a selection bias) or events that did not occur in both groups of counties (a history bias). Nevertheless, any differences you find provide stronger evidence that the program was effective than just comparing counties at the end of the campaign.

The *interrupted time series* design builds on the basic time series introduced in Chapter 4 by including an explicit intervention. As is true of any time series you must have consistent data on the outcome of interest over time; that is, investigators must have collected the data at regular intervals using the same measure and data collection method each time. You should study the data before the intervention and look for normal fluctuations or trends. By including pre-intervention data you will have controlled for maturation, since the data suggest what would have happened without the intervention. If an intervention has an immediate impact a graph will show a clear increase or decrease right after the intervention. If the change is not immediate or sharp, visual examination of a graph may be insufficient. Gradual changes in the trend (slope of the line) may be missed. The safest strategy is to ask someone experienced in interpreting time series data to help evaluate the impact.

The notation for an interrupted time series is

$$O_1 \, O_2 \, O_3 \, O_4 \, O_5 \, O_6 \, O_7 \, O_8 \, O_9 \, X \, O_{10} \, O_{11} \, O_{12} \, O_{13} \, O_{14} \, O_{15} \, O_{16} \, O_{17}$$

The design does not eliminate history; if other events coincided with the intervention you cannot eliminate those other events as affecting the outcome. From studying the post-intervention data you may infer that the effect occurred gradually. The post-intervention data may capture diminishing effects, referred to as *decay*. For example, think about weight-loss programs. A common experience is for people to lose weight on a diet and then gradually regain it.

You can combine interrupted time series and comparison group designs. The notation for the combined design is

$$O_1 \, O_2 \, O_3 \, O_4 \, O_5 \, O_6 \, O_7 \, O_8 \, O_9 \, X \, O_{10} \, O_{11} \, O_{12} \, O_{13} \, O_{14} \, O_{15} \, O_{16} \, O_{17}$$
$$\text{(experimental group)}$$
$$O_1 \, O_2 \, O_3 \, O_4 \, O_5 \, O_6 \, O_7 \, O_8 \, O_9 \quad O_{10} \, O_{11} \, O_{12} \, O_{13} \, O_{14} \, O_{15} \, O_{16} \, O_{17}$$
$$\text{(comparison group)}$$

For example, researchers evaluated a campaign to encourage drivers in a California metropolitan area to use passenger restraints for all children under the age of 3. (This was before state laws required such restraints.) The researchers identified another metropolitan area in the state that had not conducted a similar campaign. They created a time series comparing the number of automobile fatalities of young children (0 to 3 years old) in the two areas. The time series for the first area was "interrupted" by the campaign; the series for the second area was not. The investigators found differences in the trends between the two areas after the point of interruption and attributed the differences to the effects of the campaign.

Non-Experimental Designs

An experimental or quasi-experimental design may be impractical. You may examine data from before-after or cross-sectional studies, which provide limited evidence of a program's effectiveness. Cross-sectional designs collect the data on all variables at one time. Both types of non-experimental designs lack many of the characteristics of experimental and quasi-experimental designs. The before-after design (O_1 X O_2) compares the outcome prior to the intervention to the outcome after the intervention. This design tells you if the outcome changed from before the intervention to after. However, it does not tell you if the change was due to the program or to something else. Even if you find no change or a change in the direction opposite to that expected, you have to entertain the idea that without the intervention the change would have been even more negative. Commonly, cross-sectional designs cannot control for time order nor threats to internal validity.

Cross-sectional designs apply the tools used to conduct surveys and analyze data to assess program impact. The program is considered an independent variable. For example, to assess the impact of the Good Diet, Healthy Babies program, you might make county the unit of analysis. For each county in your sample you would record data on the percentage of low–birth weight babies (the dependent variable) and whether or not the county had implemented the campaign (the independent variable). Rather than selecting comparison counties you would include variables measuring relevant characteristics for each county. Using the tools we presented in Chapters 8 and 9 you would analyze the data to see if a statistical, nonrandom relationship exists between the campaign and baby birth weights. However, depending on when the data were gathered you may not know if the campaign was in place long enough to effect babies' birth weights. You also have to contend with the possibility that a critical variable affecting birth weights was not included in your data set.

Non-experimental designs do not allow you to attribute outcomes to program activities. The one group before and after study, however, is useful for detecting changes in an outcome variable from before an intervention to after. Cross-sectional studies, such as surveys, are particularly valuable for identifying and evaluating relationships among variables, especially if the study includes information on a large number of cases and several variables. You can statistically control variables to eliminate some threats to internal validity. The time order of variables usually cannot be established in cross-sectional studies since, by definition, variables are measured at one time only.

EXTERNAL VALIDITY

Thus far we have asked, "Does the intervention have the hoped-for impact?" We focused on internal validity, that is, identifying other factors that may have caused the outcome attributed to the specific application of the intervention. A follow-up question is, "Will the intervention work elsewhere?" No one study can provide definitive evidence that the intervention will achieve similar results at other times, in different locations, with a different organization, and with different participants.

We refer to the evidence that findings from one study can be generalized beyond the study as *external validity*. The evidence of external validity is valuable in deciding whether an intervention or program that investigators have found to be effective elsewhere will work in your community or agency. The study's community or your community, however, may have unique characteristics preventing the replication of the study's findings. Among unique characteristics are the following:

> *History.* Recent events may affect the reaction to an intervention. For example, news stories about food-borne diseases may motivate pregnant women to pay more attention to information about food safety.
>
> *Selection characteristics of the study participants.* (1) Findings from studies of pregnant women seen by private practices may not apply to pregnant women routinely treated in public clinics. (2) Study participants may be cooperative and attentive; a program may be less successful when participants are distracted by the demands of everyday life.

Rather than expecting a single study to demonstrate external validity, check to see if investigators have applied the same intervention at different times, at different sites, to different people, and with different staff. If the outcomes are similar you can reasonably assume that the findings will generalize to any population of interest.

DECIDING THAT A PROGRAM WORKS: THE SEARCH FOR EVIDENCE

To decide that a program is effective requires evidence from more than one research study. Any one study takes place within a specific context and this context may be integral to whether the program will work in other settings. Convincing evidence of a program's effectiveness can be developed through meta-analysis, that is, combining and analyzing the findings from studies with similar research questions. While the topic of meta-analysis is beyond the scope of this book, you can adapt some of its processes.

The purposes of meta-analysis are to identify (1) the overall effect of a program or intervention, (2) the conditions under which it works best, and (3) if the type of design is related to the effects found. The first purpose produces evidence of a program's effectiveness, the second turns your attention to the concerns of external validity, and the third addresses the concerns of internal validity. You would start by identifying relevant empirical studies. For practical reasons you may decide to focus on published studies. However, this decision can weaken the value of the meta-analysis insofar as studies that show no relationship between a program and outcome may not be published.[2] This creates a possible bias in favor of programs that work under some conditions.

You need to develop decision rules regarding the studies to focus on. You may want to consider research quality and confine yourself to studies that describe the sample, measures, and design and identify the study's limitations. Once you have read the studies and developed a strategy for weighing the evidence you may want to stop. In practice an actual meta-analysis goes further and combines the data to create common metrics. As you read the studies you should have developed some hypotheses about whether the intervention works, how well it works, and under what conditions it works.

As you review existing studies you may run across reports on pilot studies or demonstration projects. *Pilot studies* are a useful first step to learning if a proposed program works. They may be initiated to determine effectiveness, but they are most valuable in developing and documenting the implementation of the program and problems encountered. Whether the pilot study is a randomized or comparison group study it still has the limitations of any other single study, and its external validity cannot be assumed.

COST-BENEFIT ANALYSIS

After deciding that a program is effective you may want to determine its efficiency. Cost-benefit analysis (CBA) asks, "Does the program's benefits outweigh its costs?" or "Which program will achieve the greatest benefit at the lowest cost?" A related question is "How much does this program cost for every unit of service (such as a person, family, or school served). Before advocating for a program, policy makers may want answers to these questions.

After a program is implemented, CBA adds to the information decision makers may want in order to decide whether a program is worthwhile. On the surface cost-benefit analysis seems deceptively simple. Its equation is

$$\frac{\text{Sum of benefits}}{\text{sum of costs}}$$

<1 if the program's costs out weigh its benefits

$=1$ if the program breaks even

>1 if the program's benefits outweigh its costs

To compute the CBA you need to express both benefits and costs in monetary units. For social programs converting social benefits into a dollar amount is not straightforward. In preparing a CBA, U.S. federal agencies often must decide on the cost of saving a human life. How would you handle this challenge?[3] Do you

---

[2]Rosenthal, R. *Judgment Studies: Design, Analysis, and Meta-Analysis* (New York: Cambridge University Press, 1987), 223–225.

[3]U.S. agencies use value of a statistical life (VSL) to estimate the value of saving a life, but VSL does not avoid controversy. In 2008 the Environmental Protection Agency dropped the VSL by nearly $1 million in a five-year period. This change had the effect of lowering the value of environmental protection programs. To learn more see S. Bornstein, "AP Impact: A human life worth less today," posted at http://www.boston.com/news/nation/articles/2008/07/10/ap_impact_an_american_life_worth_less_today/, accessed July 10, 2008.

think that some lives should be valued more highly than others? Some decision makers put a greater value on the life of a young person than on that of an old person; some assume that a college graduate's life is worth more than a high school dropout's. To avoid the ethical dilemmas and logistic problems of monetizing outcomes you may choose cost-effectiveness analysis (CEA), which does not monetize benefits.

In addition to choosing between a CBA and a CEA you need to decide whose costs and whose benefits you will quantify, how to account for the value of money over time, what to do about intangible costs, and how adequate the findings are. Deciding whose costs and benefits belong in the equation is analogous to selecting a unit of analysis. For example, if government funds pay for a job-training program, the costs and benefits are those that accrue to the government. You would include the amount of money the government spends as costs; its benefits include the amount it receives from taxes from the trainees and the amount it saves in unemployment benefits, food stamps, or Medicaid. If you consider the costs and benefits from the trainees' perspectives, their costs may include the money they spent to enroll and attend training and a benefit might be increased income after training. Note, that savings realized by the government, such as the cost of food stamps, may be experienced as a cost by program participants.

Issues associated with operational validity are deciding: how to handle the value of money over time, how to handle intangible costs, and how adequate the findings are. To account for changes in the value of money you should to use a percent, called the *discount rate,* when calculating CBA or CEA for a multiyear project. You may select a discount rate based on practices in your agency or community; the specific rate chosen may affect whether a program is considered efficient or not. Whether you use CBA or CEA you cannot avoid the problems of how to handle intangible costs and benefits, such as stress, security, contentment, frustration, and other outcomes that social programs may produce. The decisions you make will affect the operational validity of your measures of costs and benefits and the results of your calculations.

Deciding on the cost per unit of service is easier. You do not need to choose among alternative units of analysis or monetize intangible costs, insofar as cost per unit is most likely to drive budget decisions. The appropriate unit of analysis and cost would be the actual money spent by the agency to deliver the program. You may need to include a discount rate if you plan to show how the cost per unit will vary over time.

Because of the detailed requirements of developing good cost estimates and performing CBA or CEA we have focused only on the issues that directly relate to methodological concerns. You will want to seek other sources and perhaps take other classes if you wish to conduct a cost study.[4]

---

[4]A search engine should direct you to good sources which provide details of CBA and CEA. We have found that the materials produced by the Urban Institute are complete and clear; for examples see S. Lawrence and D. P. Mears, *Benefit-Cost Analysis of SuperMax Prisons: Critical Steps and Considerations* (Washington, D.C.: The Urban Institute, 2004). Posted at http://www.urban.org/UploadedPDF/411047_Supermax.pdf, accessed February 15, 2010.

CONCLUDING OBSERVATIONS

Developing evidence that a program is effective and efficient is an ongoing process. Although experimental designs and cost-benefit analysis seem perfect for identifying programs that should be adopted, disseminated, and kept, in practice neither is easy to implement. Experimental designs rely on random assignment of subjects and the researcher's control of the research setting. Random assignment may be ethically or logistically infeasible. (Ethical questions arise if the control group is being denied a treatment that may be effective.) Furthermore, over time, the assumed similarities between the experimental and control groups may decrease as participants drop out and the fidelity of the original design is lost.

Quasi-experimental designs use comparison groups, interrupted time series, or a combination to approximate experimental designs. The greatest difference is that quasi-experiments do not include random assignment. Non-experimental designs and cross-sectional studies provide weak evidence of a program's effectiveness, but they may work well as a first pass in studying a program's effectiveness.

As politicians and program administrators seek to adopt evidence-based practices, meta-analysis and pilot studies are becoming increasingly important. Investigators use meta-analysis to compensate for the limitations of any one study. By examining similar studies investigators hope to identify common themes and trends to provide stronger evidence of how well a program or intervention works. Pilot studies are adopted to work out the logistics of implementing a program and to estimate its effectiveness outside the political spotlight.

Based on the evidence of program effectiveness and knowledge of the ratio of costs to benefits (either cost-benefit analysis or cost-effectiveness analysis) policy makers may have the information they need to decide whether to adopt, expand, or maintain a program.

RECOMMENDED RESOURCES

Sylvia, Ronald D., and Sylvia, Kathleen M, *Program Planning and Evaluation for the Public Manager*, Third Edition (Long Grove, IL: Waveland Press, 2004). This book includes a chapter on how to do a cost-benefit analysis.

Weiss, Carol H., *Evaluation: Methods for Studying Programs and Policies*, Second Edition (Upper Saddle River, NJ: Prentice Hall, 1998).

Practitioner-oriented guides and examples of program evaluations, outcome measures, and cost-benefit analysis may be found on the Urban Institute (www.urban.org) Web site. The United Way Web site (http://www.liveunited.org/Outcomes/Library/pgmomres.cfm) lists resources for outcome measurement.

CHAPTER 11 EXERCISES

Exercises in this chapter are presented in two parts. In the first section the exercises focus on the methodology of research designs used for evaluation. In the second section we review tests of statistical significance in the context of program evaluation and include some follow-up questions.

- Exercise 11.1 An Antiflu Campaign uses a previous exercise (Exercise 4.1) on an antiflu campaign. The exercise asks you to apply different evaluation strategies: general approaches to evaluation, randomized designs, quasi-experimental designs, cross-sectional designs, and cost analyses.
- Exercise 11.2 Prenatal Support Program asks you to assess data a women's center has collected to evaluate its program to reduce the incidence of low–birth weight babies.
- Exercise 11.3 On Your Own asks you to describe the components of an evaluation done for your agency and to learn how its findings were used. You are also given the opportunity to identify a program that your agency would like evaluated and to recommend an initial design.

EXERCISE 11.1 An Antiflu Campaign

Scenario 1

In Chapter 4, Exercise 4.1 you were asked to create a logic model for the Mid Valley Family Health Center. The center is launching an aggressive antiflu campaign in the communities served by its clinics. The campaign will consist of two major activities: a publicity campaign about behaviors that can prevent the spread of flu and providing flu shots at no cost. The clinics, which serve low-income households, are located in racially and ethnically diverse communities

Section A: Getting Started

1. What is the difference between a process evaluation and an outcome evaluation, the two major approaches to evaluating a program?
2. What is the value of starting with a logic model in
   a. designing a process evaluation?
   b. designing an outcome evaluation?
3. Write three process evaluation questions that might interest the health center director.
4. Write three process evaluation questions that might interest a donor who was paying for fliers and public service announcements.
5. What outcome measure(s) would you use to
   a. evaluate the publicity campaign?
   b. evaluate the flu clinics?

Scenario 2

Prior to launching the publicity campaign a video is produced. Investigators want to learn if it effectively communicates accurate knowledge about influenza and seasonal flu shots. Investigators will assign people to watch one of the two videos: the new video (NEW) and a video from a newscast on flu (OLD). After viewing the video, subjects will be given a short questionnaire asking questions about the flu and flu shots.

Section A: Getting Started

1. The investigators consider conducting the above study with (a) people who have been called for jury duty and are waiting in the jurors' lounge, randomly assigning people to a Website where they will watch either NEW or OLD, or (b) people attending movies at a multiplex theater with 12 screens; 6 theaters will be randomly assigned to show OLD and the other 6 will show NEW.
   a. Would you use a pretest-posttest design or posttest only design? Justify your answer.

b. Compare the two proposed designs for the following threats to validity: (i) history, (ii) maturation, (iii) selection, (iv) experimental mortality, and (v) design contamination.

Scenario 3

Assume that the investigators have monthly data on the number of flu cases and the flu rate for your community and for specific age groups. Statewide data are available for cities over 10,000 population.

Section A: Getting Started

1. Would you use one year's worth of data or several years' worth in an interrupted time series? Explain your choice.
2. To create an interrupted time series with a comparison group
   a. suggest a community that you would use as a comparison group. Explain what characteristics made this an appropriate choice.
   b. explain why an interrupted time series with a comparison group is an improvement over the interrupted time series.

Scenario 4

A month after the last clinic a random sample of health center patients is surveyed. The database contains the information for each individual.

- Did you receive a flu shot this September? (yes/no)
- This year were you hospitalized with flu? (yes/no)
- This year were you diagnosed with flu, but not hospitalized? (yes/no)
- This year did you have flu symptoms but didn't receive medical care? (yes/no)
- Do you regularly get a flu shot? (yes/no)
- Did you see ads about the value of getting a flu shot (yes/no/don't remember)
- Were you urged by family or friends to get a flu shot (yes/no)
- Age
- Census track (filled in by staff)

Section A: Getting Started

1. You decide to analyze this data using a cross-sectional study.
   a. Identify your independent and dependent variables.
   b. Briefly describe how you would analyze the data.
   c. If the analysis showed that persons who got the flu shot were less likely to (i) be hospitalized with flu, (ii) be diagnosed with flu, or (iii) have flu symptoms, consider each finding separately (i, ii, and iii) and discuss what it suggests about the effectiveness of the flu shot that the flu shots were effective.
2. The center decides to do a cost-benefit (or cost-effectiveness analysis) of both the publicity campaign and the flu clinics. For each campaign identify
   a. the costs you would include.
   b. the benefit you would track.
   c. whether you would recommend CBA or CEA. Justify your choice.

Section B: Class Exercise

1. Working in small groups design a study to evaluate the antiflu campaign and prepare a presentation. In your presentation
   a. identify the question(s) your study should answer.
   b. use the R, X, O notation to illustrate your design and indicate
      i. what X represents.
      ii. what O represents and how you plan to measure it.
      iii. how you are going to select subjects (use R only if you are going to have random assignment).
   c. Identify the strengths and weaknesses of your design
2. Groups that have chosen different designs will present their designs to the class. The class will discuss and evaluate each group's design. It should rate each design's feasibility, quality, and ease of implementation.
3. Based on the presentations, the group discussions, and the Getting Started questions the class should develop a lessons learned list. The list could be divided into the value of a logic model, information needed before creating a research design, and criteria for selecting a research design.
4. The class uses what its members found in "getting started" to develop evidence of the effectiveness of publicity campaigns.
   a. What are some sources or search strategies for locating relevant research?
   b. What research models were predominant?
   c. What criteria could be used to decide between research that should be included and research that can be ignored?
   d. What evidence was found that suggests effective and ineffective campaign strategies?

EXERCISE 11.2 Prenatal Support Program

Scenario 1

Rain County's Women's Center created a program to reduce the high percentage of low-weight births. The program consists of nutrition classes and transportation for women and their partners to prenatal clinics.

Section A: Getting Started

1. The staff has birth weight data for the year prior to the initiation of the program and for the year ending 3 years after the program started. The first-year data showed that 9.5 percent of babies were low–birth weight; the end of the third-year data showed 8.1 percent were low–birth weight. The staff member wants to use this information as evidence of the program's effectiveness.
   a. What type of design has the staff member suggested? Diagram it with the R, X, O symbols.
   b. Critique the design. Use the threats to internal validity and the evidence necessary to demonstrate a causal relationship to evaluate whether the evidence supports the conclusion that the program was effective.

2. The Rain County Health Department has the following annual data on the percentage of low–birth weight babies born in the county

Year	Percentage Low Birth-Weight Babies
1	10.5
2	9.8
3	9.2
4	9.2
5	9.5
6	9.5
7	9.2
8	8.3
9	8.1
10	8.0
11	8.2
12	9.1
13	9.1
14	8.5
15	8.0

a. Draw a time series graph using these data. Assume that the women's center program was implemented at the end of year 6.
b. Suppose the women's center's staff claim that its program was effective. Would you say that the graph supports their claim? (You may want to review Chapter 4 to refresh your understanding of time series graphs.)

Section B: Class Exercise

1. A national organization has requested proposals for innovative programs to reduce the percentage of low–birth weight babies. You have been assigned the task of describing the pilot study of a program similar to the women center's program.
a. Explain what a pilot study is.
b. What information can a pilot study provide to policy makers?
c. Develop a design for the proposed study.
  i. Describe how subjects would be selected.
  ii. What data would you collect? When would you collect it? How would you collect it?
  iii. What are the strengths of your design?
  iv. What problems might you anticipate if you implement the design?
d. In your opinion how long should a pilot program last before its effectiveness is determined? one year? two years? three years? longer? Justify your choice.
e. If the program is funded and later an evaluation produces evidence that it is effective, identify factors that should be considered before recommending widespread dissemination of the program.

EXERCISE 11.3   On Your Own

1. Suppose that the agency you are working in has evaluated its programs within the last 3 years.
   a. What questions did the evaluation ask?
   b. What research design was used?
   c. What measures were used? Did the evaluators cite evidence that the measures were reliable and valid?
   d. How were the data analyzed?
   e. What were the major findings?
   f. Describe any actions the agency took based the evaluation findings.
   g. Ask staff to describe the evaluation process and the usefulness of the findings. What features of the evaluation process or findings motivated the agency to take action or make a decision?
2. Suppose that the agency has a program it wants evaluated.
   a. Draw a logic model.
   b. Meet with the appropriate program staff to explore whether a process or outcome evaluation would be better.
   c. Identify what data are available or routinely collected that might be used in an evaluation.
   d. Suggest a design that the agency could use to evaluate the program.
      i. Sketch out the design using R X O notation.
      ii. Explain how the agency would select subjects, introduce the intervention, and determine what data they would collect and when and how they would collect it.

APPENDIX TO CHAPTER 11

Analyzing Experimental Findings: Discussion and Questions

To analyze evaluation data you may use statistics that we introduced in Chapters 5, 8, and 9. In this section we illustrate three statistics commonly included in evaluation reports; they are relative frequencies, means, and tests of statistical significance. Recall that tests of statistical significance help determine if a relationship found in the data was probably due to chance.

To compare data from different groups you may decide to compare relative frequencies or means. For example, to study pregnant women's dietary knowledge, women were given nutritional information in two different formats. Each woman was asked whether she should "eat" or "pass" on each of four menu items. Each menu included some items that were safe to eat during pregnancy and others were not. Table 11.1 shows the percentage of women in each group that got 4, 3, or fewer than 3 correct answers.

You could analyze the data by examining means; the choice is yours. Table 11.2 shows the average number of correct answers given by the women in each group and the standard deviation for each group. You can analyze these figures to determine if using the new format increased what the women learned over the existing format.

TABLE 11.1

Knowledge of Safe Foods by Formats Received

Number of Correct Answers	New Format (Experimental Group) ($n = 39$)	Control Group (Existing Format) ($n = 39$)
4 out of 4	69%	33%
3 out of 4	13	21
Fewer than 3	18	46

TABLE 11.2

Mean Number of Correct Answers by Formats Received

	New Format ($n = 39$)	Existing Format ($n = 39$)
Mean	3.0	2.1
Std dev	1.06	1.13

In reviewing either table you may want to ask the following:

How large is the difference between the two groups?
Is it possible that the difference between the groups occurred by chance?

The first question involves effect size. Effect size tells you how strong the relationship is between an intervention and an outcome. More precisely, it yields a metric that indicates how much of a difference the program intervention made. To determine effect size you may compare the differences in means or percentage differences between the two groups.

The second question is associated with statistical significance. Two common tests of statistical significance are the χ^2 (chi-square) test and the t-test. The t-test is based on each group's mean and standard deviation.

A test of statistical significance asks, "Could the observed relationship have occurred if the relationship did not exist in the population?" In experiments, that question may be stated as, "Are the differences among members in a group less than the differences between groups?" In both cases a finding of significance implies that the difference between groups did not occur by chance. If we found a relationship was "statistically significant" we may infer that an intervention had an impact.

For tables similar to Table 11.1, the common significance test is χ^2. The χ^2 value for Table 11.1 is 10.4 and the p value is $<.001$. The probability (p) indicates the probability that in the population the percentages of the two groups are similar. $p < .001$ indicates a less than 0.1 percent chance that in the population the two groups are similar. Another way of thinking about the information is that the probability is less than 0.1 percent that the observed relationship occurred by chance. In this case, if we are relatively certain that the intervention was implemented as designed and threats to validity were controlled, we may conclude that the program had an effect. In practice p values of $< .05$, $< .01$, and $< .001$ are considered significant. Which p value is chosen depends on the size of the sample and the effect size desired. A common practice is to use $p < .05$ to decide that a relationship is statistically significant.

So what do you want to remember when looking at experimental data? First, the important information communicated by χ^2 or t is the value of p. (Statistical tables or software identify the p value associated with the given χ^2 or t value.) Second, a finding of statistical significance eliminates random error, or chance, as a rival hypothesis. To eliminate other rival hypotheses one has to see how well the research design as implemented controlled for threats to internal validity. Third, the strength of the relationship is measured by effect size and not by the value of χ^2 or t.

Ask the same questions when you are interpreting a randomized pretest-posttest design. For effect size the change between the pretest and posttest for both the experimental and control groups are compared. For statistical significance the size of the change between pretest and posttest can be computed. You would hope to find that the difference between the pretest and the posttest is significant for the experimental group and not for the control group. Or you would expect to find that the size of the change for the experimental group is significantly different from the size of the change for the control group. Statistical significance may be used to verify that the experimental and control groups are equivalent at the beginning of the study.

QUESTIONS

1. In Table 11.1 how large is the difference between the percentage of women in the two groups who answered all questions correctly? In Table 11.2, on average how many more questions did the experimental group answer correctly than the control group?

2. Do you think that these differences are large enough to conclude that the new format was better than the existing format? What else would you want to know before making this decision?

   Note: Deciding whether an observed effect is large, small, or somewhere in between is a matter of judgment and experience. For example, health educators may have sufficient experience to decide how much difference justifies changing existing materials.

3. Table 11.2 has a t-value of 3.6 with $p < .001$. Explain what this means. Would you conclude using the t-test that the program was successful?

4. Use a search engine to find an article reporting an evaluation and applying either χ^2 or a t-test. How did they report the value of test statistic (χ^2 or t) and the probability level? Was an effect size also reported? If means were reported were confidence levels also reported?

Community Needs Assessment

A s planners, policy makers, and practitioners, we are often in a hurry. We may feel compelled or under pressure to find and implement a solution to a social problem, but we do not always select the best intervention or correctly identify the problem. To ensure that implemented programs meet an existing need, agencies may conduct a needs assessment. A needs assessment develops systematic, credible information that

- identifies the need(s) of interest;
- describes who has the needs, that is, the target populations;
- documents the availability of services and programs to alleviate the needs.

An agency that conducts a needs assessment avoids wasting human or financial capital if for instance it has to decide what to do with a major gift or whether to apply for a grant. A needs assessment may also be called for if an evaluation indicates that a program is not working. To identify needs, who has them, and what services are available, an agency may conduct surveys, create resource inventories, analyze service statistics, consult existing studies, conduct interviews and focus groups, or hold community forums.

IDENTIFYING NEEDS

You cannot start a needs assessment with no idea of what needs to study. Individuals and communities have numerous needs and you would soon be overwhelmed. Besides, an agency with an education focus may not want to explicitly study health needs that it is in no position to address. To focus the needs assessment, a management team or community task force may draw on the agency's mission statement or objectives. Consider the following objectives:

To improve the local economy in [location]

To decrease infant mortality among [population group]

To increase the supply of affordable housing in [location]

TABLE 12.1

Comparison of Wants and Needs

Want	Need
To attract more businesses	To improve the local economy
To improve maternal health care	To decrease infant mortality
To build affordable housing	To reduce homelessness

Once you have identified the area of need (local economy, infant mortality, or substandard housing) you will want to consider how to identify specific needs. As you begin to identify needs you may want to distinguish them from wants. *Needs* are gaps, that is, the space between what currently exists and what could or should exist. *Wants* are solutions or a proposed strategy for filling a gap. Table 12.1 provides examples of needs versus wants.

While you need not split hairs distinguishing needs and wants, you may want to ask, "What problem are we trying to solve? for whom?" The answers may focus managers' attention and help them arrive at better solutions. Let's consider homelessness. People are homeless for many reasons, including unemployment and chronic mental illness. The appropriate solutions for the unemployed might include job training and job placement along with temporary housing. For persons with mental illness, however, supportive housing, which provides mental health services, may better reduce homelessness. These solutions do not eliminate the need for affordable housing. We know that housing may be out of reach of persons who earn minimum wage. The example illustrates the benefit of not embracing the solution—what is desired before understanding the need.

Concept of Need

In designing a needs assessment we need to clarify and agree what is meant by "needs." As stated above, needs are gaps between what exists and what should exist. Despite this straightforward definition precisely defining *need* is a difficult task. Therefore in planning, program development, and policy making the word need is commonly used but rarely defined. Perceptions of need are influenced by multiple social-cultural-political factors. Bradshaw (see Recommended Resources), a pioneer in the contemporary conceptualizations of need, developed a typology of need that included perceived, expressed, normative, and comparative needs. We discuss each briefly below.

Perceived (Felt) Need: A perceived (or felt) need is anything people consciously lack and desire. This type of need is a need that people feel, in other words, it is a need from the perspective of the people who have it. The following hypothetical needs of individuals, apartment residents, and welfare mothers illustrate perceived needs. Individuals may feel they need better street lighting; apartment residents may feel they need more parking spaces rather than coming home from work and having to walk to their building; or welfare mothers may feel they need affordable quality child care, especially if they are working a 2nd shift. Perceived needs may be influenced by the awareness and knowledge about what could be available. For example, mothers may not feel a need for high quality child care until they visit a state of the art child care center.

A perceived need may not be adequately expressed because potential users may not be aware that the need can be remedied. Additionally, a perceived need may not be articulated because existing solutions are deemed inadequate and therefore irrelevant. For instance, historically, in the African American community, even though individuals may have experienced significant emotional and mental distress, mental health therapy was considered unacceptable. Thus, community members may not have felt a need for these services. Perceived need is the extent to which people think their needs are being met. Perceived need is quite elastic and unstable.

Expressed Need: An expressed need is a perceived need that is given voice. Expressed needs come in the form of a demand or request. Logically, perceived needs may become expressed needs. However, expression is not always an option. Due to the lack of power, resources, and conduit, many of the perceived needs of individuals, such as low-income residents or immigrants, go unvoiced. There may be a reluctance to express needs for community improvement due to perceived retaliation from policy makers. Often, policy and decision makers assume that needs that are not expressed are not needs at all. Expressed needs can, however, be misinterpreted. Long waiting lists at a health service may be the result of inefficiency and not the size of the group wanting treatment. Additionally, a need may exist but because no service is available to meet it, the expressed need may not be identified.

Normative Need: Normative needs are based on a professional assessment or knowledge. A normative need is defined by an expert, a professional, a policy maker, or a social scientist according to established standards; falling short of those standards means that a need exists. Poverty thresholds are an example of normative needs. Normative need is often considered unbiased and objective; however, it is influenced by the social standing, position, values, and biases of the experts. Further, normative needs are influenced by political and other agendas. This lack of objectivity is reflected by the fact that normative needs change over time. For example, Departments of Social Services may require that case workers inquire about domestic violence when interviewing clients. Until recently, protection from domestic violence was not considered a need. Not until experts identified the high incidences of domestic violence among social service recipients was this screening necessary.

Comparative Need: A comparative need (also called a relative need) is based on others who receive a service. A person is in comparative need if he is not receiving a service that someone with the same or similar condition is receiving. An example is the perceived need for community-based banking in one neighborhood because it was offered in a neighborhood with similar residents. The problem with comparative need is that it has as a premise that needs are the same or similarities are close enough to be adequate predictors of need. Comparative need is derived from examining services provided in one area to one population and using this information as the basis to determine the services required in another area with a similar population. When assessing comparative need, the level of service provided in the reference area must be appropriate in the first place. Be aware that data collected may reflect overservicing or underservicing by service providers rather than an indication of true need for the service by service users.

Conflicting Needs: While Bradshaw's typology identifies perceived, expressed, normative, and comparative needs, it is worth mentioning a fifth need, conflicting needs. Any one of the needs discussed thus far taken alone may not be adequate to assess the actual needs of a community and its residents. However, addressing the needs in concert allows for a more robust and well-developed assessment of need. As community needs assessments are conducted, the voices of multiple constituent groups (residents, professionals, leaders) must be heard to sift out the actual need. Each of those constituent groups may have different needs that are not complimentary to one another. This circumstance is particularly relevant when considering resources. Often resources are not available to address multiple needs. For example, if half of the survey responses suggest that community members want a recreation center and the other half want a community garden, the conflict occurs when trying to determine whose need takes precedence or is addressed.

DESCRIBING NEEDS AND SERVICES

You need to generally know what type of need that you want to assess so that you understand the implications of your findings. Still as you start your needs assessment you do not have to get bogged down trying to define the specific type of need. Be aware that no single best way of assessing the needs of a particular community or group exists. The methods that you choose will completely depend upon your circumstances. In embarking on a needs assessment some important questions to ask during planning include,

- Who is the assessment attempting to inform, influence, or persuade? Who will be the target audience for the needs assessment report?
- What is the purpose of conducting the needs assessment? What is the intended change or outcome that is desired?
- Whose needs are to be assessed? Who is the target population?
- What questions should be asked? Do you already know the answers? What sources of data can you access to get the answers? Once you ask the questions, can you do anything to change the situation?
- How will the information be used?
- What human, financial, technological resources are available to conduct the needs assessment?

Having clarity about the answers to these questions will help you determine the best course of action and way to proceed. Answering these questions will help set the parameters for a needs assessment and keep you from wasting time and precious resources. So, before beginning to conduct a needs assessment, you should be clear about what you want to measure, and for whom you want to measure it. The more specific and focused your questions, the better you can choose the best method for answering the question.

Think about needs assessment as an iterative process. As you work with a data source you are both trying to better define the needs of interest and to describe the population. You may find that as you better understand the need, you will go back to some data sources to better describe the population. In general we recommend starting with existing data sources since these may increase the efficiency of later surveys, interviews, or community forums. Existing data sources are the easiest to revisit if you want to add new information. On the other hand the data may be less specific than you hoped for and they may only serve as a beginning point.

Collecting Information from Databases and Agencies

Secondary Data: Secondary data are existing data that have been put into a database. They will help you identify needs and describe the population that has those needs. Databases may contain information on standard indicators of a community's well-being, such as infant mortality or percent of population that has a high school education; labor force data; data on program budgets; and data on characteristics of the population receiving services. The U.S. Federal Government, especially the U.S. Census Bureau, is a major source of secondary data (to identify available data go to www.fedstats.gov). Libraries and research centers have developed archives and links to databases: one source is the Cornell Institute for Social and Economic Research (http://ciser.cornell.edu). Relevant data may also be found in annual reports, planning and budget documents, and personnel evaluations.

Let's now consider population data. A strategic manager can use data on the variations in population to identify changing needs for schools, police, and other public services. Areas experiencing rapid growth may have inadequate services, and areas with declining growth may have to cut back on existing services. But this observation may not tell the whole story. How did the population change? Were there fewer young people? Did the immigrant population increase? Answers to such questions will yield data that will uncover more about changing needs than just tracking changes in population size.

Working with existing data saves time and money and avoids burdening respondents. The data can be obtained and analyzed relatively quickly. Many social statistics and social indicators are collected regularly and form a time series. Consequently, they can be tracked over time, and may identify normative needs and comparative needs. Databases maintained by polling organizations may have information on expressed needs. You will rarely come across databases that identify felt needs. You may find public-opinion databases, but they may be inadequate to identify local needs.

When working with secondary data you must pay attention to details. First, a measure may be broader than you need. For example, it may provide state data when you want to identify local needs. It may tell you how many people are disabled but it may not tell you how many have specific physical or mental disabilities. Second, if you are comparing needs, raw numbers may be misleading. Rather, you should consider the population size and calculate the rate. You also need to make sure that you selected the appropriate denominator. Sometimes this is difficult. Consider the mortality rate for a strain of influenza. The two communities you may be comparing may differ in important ways. In one community, people may be more likely to seek medical attention than in the other. Medical personnel in one may be more likely to run tests to confirm and report the flu strain than those in the other community.

A related issue is how variables have been measured. Their operational definitions may differ from study to study. A given definition may not be appropriate for your purposes. For example, if you plan to develop a program that addresses child sexual abuse, you have to define what you mean by *child sexual abuse* before you can measure its prevalence. Here are three different, definitions of the term:

- Any contact or action initiated by an adult towards a child, which is intended to stimulate the child sexually, or to stimulate the offending adult through the use of the child. (Family Service Moncton, http://www.fsmoncton.com/eng/services/child_sexual_abuse.htm)
- An act imposed on a child who lacks emotional, maturational, and cognitive development. The ability to lure a child into a sexual relationship is based upon the all-powerful and dominant position of the adult or older adolescent perpetrator, which is in sharp contrast to the child's age, dependency, and subordinate position. (SusanSmiles.com, susansmiles.com/lynne.html)
- Inappropriately exposing or subjecting a child to sexual contact, activity, or behavior. Sexual abuse includes oral, anal, genital, buttock, and breast contact. (TeenhealthFX, http://www.teenhealthfx.com/answers/Emotional/4740.html)

You must determine if one of the definitions is consistent with your definition or how you plan to interpret the data. If not, you will need to develop your own definition.

Existing records, such as client files, that are not entered into a database may have valuable information. However, compiling the information is time consuming. Also, you need to be aware of legal and ethical constraints that may apply if you want to review records that may for instance include confidential information such as performance evaluations. Again, you need to consider the reliability of the information and make sure the data collectors were well trained and supervised while collecting data. The directions on what information to collect and how to record it must have been clear.

Resource Inventories and Service Statistics: A resource inventory lists available services. You may have to refine an existing list and confine it to services available to your target population. Or you may have to collect information to create your own list. For example, chronic welfare recipients typically need food; stable housing; mental health, medical, and dental services; emergency financial assistance; job training and placement; substance abuse treatment; parenting education; and case management. A resource inventory for a city may list all of the city's public, private, and nonprofit agencies that serve welfare recipients. To limit the list to services for welfare recipients, you would survey the listed agencies to ask if they provide specific services. You might also ask about the availability of services, such as hours of operation, access by public transportation, and typical waiting period before receiving services. To determine who uses the services you may collect data when a person first requests services, such as address, gender, race, or ethnic group. Once you have collected and aggregated this information you would be better able to decide if available services are adequate or where needs persist. You will also have identified areas with over-capacity or under-capacity and duplication of services.

Resource inventory data are collected from experts (program directors and other administrators) rather than the actual clients. Consequently, the inventories reflect normative needs rather than perceived needs of clients or nonusers. For example, clients may not come to an agency because they are unaware of the service, they consider the services unacceptable, or they have heard that the agency has long wait times for services. Therefore, such data may not adequately reflect needs.

Collecting Information from Individuals

Individuals are an important source of information on perceived and expressed needs and on the quality and availability of services. You may survey individuals, conduct intensive interviews, or hold focus groups. Since you want to generalize from the findings you cannot ignore sampling issues. It is particularly important to assess the importance of sample error and sample bias in your findings. In addition to the number of people you will sample, you need to consider salient characteristics that affect the representativeness of your sample. Getting a sufficient number and diverse sample of community members or key stakeholders may require invitations, incentives, follow-up calls, or even door-to-door canvassing.

Surveys and Interviews: Surveys are appropriate for gathering data on needs and services. You can conduct a survey for a needs analysis applying the information in Chapters 5 through 9, which covered writing questions, administering the survey, and analyzing the findings. In Chapter 7 we included qualitative interviews under the topic open-ended questions. For the most part our discussion applied to structured interviews, where each person is asked the same questions in the same way. On the other hand, needs assessments may use unstructured interviews, which may be particularly valuable in identifying perceived, unexpressed needs.

Unstructured interviews are conducted with key informants selected because of their unique perspective. When conducting an unstructured interview, the interviewer wants to create an atmosphere and a process that encourage respondents to share their knowledge and opinions. Interviewers try to promote candid, complete responses by asking questions in a different order, asking different questions, or allowing respondents to go off on a tangent. Some investigators call this approach *responsive interviewing*, which alludes to the fact that the interview proceeds based on the responses of the participant.

Responsive interviewing requires practice, preparation, and skill. The interviewer must be a good listener, comfortable with interpersonal relationships, able to engage respondents with thoughtful questions and appropriate probes (discussed in Chapter 7, in the section Questions for Interviews and Focus Groups). The skills of active listening, observation, and critical thinking are required.

To identify perceived needs and learn more about the quality and accessibility of services you may want to interview clients or potential clients. Potential interview subjects could include mentally ill persons, released prisoners, or pregnant teenagers. To learn more about normative needs and the service delivery system, including barriers to creating and providing services, you may want to interview service providers, agency heads, politicians, or researchers. The first set of groups listed above may be wary. The more structured the interview, the less likely you are to tap into feelings associated with psychiatrists, parole officers, and case workers, and more likely to get wooden responses. With unstructured interviews, if you can gain a respondent's confidence and adapt your questions in reaction to her answers, you are most likely to be rewarded with valuable information.

At this juncture, it is worth taking a moment to revisit our discussion on ethics (remember Chapter 3). While we may be able to glean substantive and important information from vulnerable populations such as mentally ill persons, released prisoners, or pregnant teenagers, we have to be especially concerned about making sure that we are protecting such individuals from harm during the research process. It is particularly necessary to gain informed consent and avoid even the appearance of coercion when working with populations that have significantly lopsided power relationships with the agencies with which they are associated. For instance, prisoners may feel like they have no choice but to participate in an interview—even if you tell them otherwise. Therefore, it is essential that informants be chosen carefully and purposefully.

Interviews with knowledgeable informants may give you insights into normative needs and help you build a case to support a need assessment's recommendations. You may hear opinions about the quality of available services, what could be done to improve or expand existing services, and major barriers to introducing new services. You can tailor your questions to take advantage of a subject's knowledge instead of coping with the one-questionnaire-fits-all approach of structured interviews. Prior to interviewing with knowledgeable informants, find whatever relevant information you can about the person you are interviewing. This will keep you from asking questions that are redundant. You would want to make sure that you are asking questions based on the participant's special knowledge. After all, that is why you are interviewing this person! The informants may have information posted on the Web, which you can access using a conventional search engine. Asking questions that follow up on, rather than repeat, publicly available information shows that you have done your homework, that you are prepared, and that you want to learn from your informant and not waste her time.

Once you have decided whom to interview, give some thought to how you will initiate the interview. You may have to interact with *gatekeepers,* those individuals who manage or control access to participants. Gatekeepers may be executive assistants, receptionists, or legislative aides. A gatekeeper plays a valuable role and can make your interview happen or prevent it. Furthermore, a gatekeeper can influence how a respondent views you. The gatekeeper may report that you are trustworthy, knowledgeable, worth talking to, clueless, or annoying.

Whether your first contact is with a gatekeeper or the intended participant, remember this person is doing you a favor, so phrase your request accordingly. Start your request by identifying yourself, the purpose of the interview, and why you would like to interview the particular stakeholder. As part of arranging the interview, suggest that the respondent choose the time and place and indicate how much time the interview will take.

Be sure to decide how you will record the interview. Audio recordings are common and easy to work with. As we mentioned in Chapter 10 many experienced researchers continue to use audio tapes as a back-up to newer digital recordings due to the unreliable nature of technology. Consider using a small recorder with a microphone that is sensitive enough to capture a normal conversation, and remember to have extra batteries and tapes. While you do not want to take detailed notes, because you will be actively listening, you still want to jot down thoughts and topics you may want to follow up on. Now you are ready to go to the interview!

At the interview, observe your surroundings. One of the benefits of qualitative research is the ability to observe context. Take the time to observe the atmosphere. Is it formal or informal? Is it orderly? How are people interacting with the space? Are they comfortable? Are there hierarchal cues? Does the furniture promote interaction or serve as a boundary?

Before you begin the interview explain any consent forms, have them signed, and ask for permission to tape record the interview. However, if you make a big deal of consent forms and tape recorders, you will turn the focus away from the research you are conducting. Before you leave, thank the participant; you should also follow up with a thank-you note. Remember most participants are doing you a favor and taking time from their day to talk with you. As soon as you can, write up your notes while they're fresh in your mind. Jot down any immediate impressions you have that may contribute to your understanding of the interview.

Focus Groups: Focus groups are semi-structured group interviews. A focus group may consist of clients or potential clients, service providers or potential service providers. Whether they are clients or service providers, participants may be impacted if the recommendations of a needs assessment are implemented, so their perceptions of needs and services may be particularly valuable.

Generally, the investigator invites focus group participants to meet at a community location. The room should be set up to encourage easy interactions among participants. The conversation of the focus group is tape recorded and later transcribed. Because the facilitator has to be fully engaged, use of audio recorders is especially important. Running out of tapes, needing batteries or other power sources, or coping with a broken recorder will be very costly, and critical information will be missed. So, more than ever it is important to have backup tapes, power sources, and an extra recorder on hand. Similar to responsive interviews you need to get consent for the taping, but don't make it seem like a big deal and draw attention to it. If possible you may want to have an assistant take notes.

Focus groups allow people to tell their stories in their own words. The group interaction adds to the richness of the conversation. One person might say something that "jogs" another person's memory or gets others to think about the situation differently. The interaction of it all is the beauty of this method. A focus group requires a skilled facilitator who can stay on task without stifling the discussion and can manage all the personalities in the room. Let's imagine some questions that might be asked of released prisoners who have found jobs. (We might ask similar questions of released prisoners who have not yet found jobs, but the conversations may be more productive if they are in a different focus group.)

■ When you were released from prison how did you go about looking for a job? (What resources did you have available? What resources did you use? What other resources could you have used?)

- When did you tell potential employers about your prison record? How did you tell them?
- In reflecting on your job-hunting experience, what did you learn that would help other men (women) who have been recently released?
- What do you think helped you get your present job? Has your prison record affected how you are treated on the job?

The facilitator may have to restrain talkers and draw out non-talkers while keeping the discussion on track (often using a topic guide) without stifling important perspectives. This task is often quite difficult as topics that seem tangential may produce gems of information. Focus groups should be homogenous to increase efficiency. Homogeneity will also decrease the likelihood of getting bogged down on issues associated with markedly different life opportunities; however, the facilitator has to anticipate and respond to the possibility of groupthink where the participants converge to verbalize the same opinions. The bottom line—a facilitator has to maintain a delicate balance between competing tendencies.

Community Forums: Community forums bring residents and stakeholders together. They discuss the needs facing the community, identify their priorities, and brainstorm solutions. All members of the community are encouraged to attend and to express their concerns pertaining to their community's well-being and needs. Community forums are often held during hours when most community members can attend, and on neutral territory, often a community facility. Nevertheless, it often happens that there is no time or day that all community members will find convenient. Often those who feel the strongest about the subject will participate and they may be very vocal. Investigators must be aware that non-attenders and quiet attenders may have different points of view that should also be taken into account. Despite these apparent limitations a community forum should attract a diverse, representative group of participants. A community forum is low cost, requires a relatively low time investment, and provides an opportunity to gather qualitative data from a cross-section of the community. Participants who are interested in participating further can be engaged on the spot, keeping the momentum going.

CONCLUDING OBSERVATIONS

Research on programs and public policy can be enhanced by the interactions among investigators. A study conducted by one investigator will be limited. Nowhere is the need for multiple researchers more apparent than in conducting a needs assessment. To identify needs, describe who has the needs, and document the availability of services and programs require accessing different sources of information using different research skills. Along the way, critical information may be unavailable; differing, and even conflicting opinions, will be heard; and the capacity to address the need may shift. To conduct a good needs assessment a plan is developed and is evaluated as new information is received. Similar to evidence-based research, the evidence about the degree of need and the appropriateness of services may not be conclusive, but the evidence should eliminate unnecessary services and support those that appear to be effective.

RECOMMENDED RESOURCES

Billings, J. R., and Cowley, S., Approaches to community needs assessment: A literature review, *Journal of Advanced Nursing* (2008), 22(4):721–730.

Bradshaw, J. R., The concept of social need, *New Society* (1972), 496(19):640–643.

Gupta, K., Sleezer, C., and Russ-Eft, D. F., *A Practical Guide to Needs Assessment* (San Fransico, CA: Pheiffer, 2006).

Witkin, B. R., and Altschuld, J. W., *Planning and Conducting Needs Assessments: A Practical Guide* (Thousand Oaks, CA. Sage Publications, 1995).

CHAPTER 12   EXERCISES

This chapter has three exercises, which take you through the process of designing needs assessment

- Exercise 12.1 Sustainable South Bronx asks you to explain the terms associated with needs assessments and to work with classmates to plan and role-play a focus group as part of a needs assessment.
- Exercise 12.2 Learning from Practice—How a Needs Assessment Looks asks you to find a needs assessment and see how the participants applied the concepts and strategies discussed in this chapter.
- Exercise 12.3 On Your Own asks you to interview someone in your agency or community who has conducted a needs assessment.

EXERCISE 12.1   Sustainable South Bronx

Scenario

Sustainable South Bronx (SSBx), a nonprofit organization, seeks to address urban policy and planning issues within the South Bronx community by using economic justice principles. The latest initiative of the organization has set out to address the needs of the Hunts Point neighborhood in the South Bronx.

The Hunts Point neighborhood in the South Bronx is one of New York City's last remaining industrial areas. On the one hand, the neighborhood has numerous assets, including a waterfront location on the Bronx and East Rivers, proximity to Manhattan, the economic engine of the Hunts Point Food Distribution Center (the second largest in the world), new City-led development projects, waterfront parks, and a strong local organizational infrastructure. Simultaneously, it exhibits one of the highest poverty and unemployment levels in the City, with poor community health, noxious uses and commercial traffic, substance abuse, and prostitution issues.

Caught in the middle of these pressures are approximately 11,000 residents who have been neglected, underserved by the neighborhood's local economy. The one-square-mile area of Hunts Point is bound by the Bruckner Expressway to the north and west, and the Bronx and East Rivers to the south and east (from http://www.ssbx.org/index.php?link=2#history, accessed December 16, 2009)

The leadership has set out to "green the ghetto." Greening the ghetto is how it describes its desire to use ecologically friendly, environmentally sound, green solutions to community problems. To demonstrate how to plan a community needs assessment, this exercise asks you to plan a needs assessment for SSBx.

Hunts Point corresponds with the following Census tracts: 31, 35, 37, 73, 79, 83, 85, 87, 89, 91, 97, 99, 105, 115, 119, 121, 127, 129.

Section B: Small Group

1. In groups of four to five, plan a focus group for the Hunts Point neighborhood residents. Each group should develop a plan to
   a. engage the residents in the efforts to green their neighborhood;
   b. conduct a focus group making sure to include (i) information about the location, time, and day for the focus group, (ii) the personnel needed to conduct the group, (iii) a sample design and recruitment strategies (refer to Chapter 6 if needed), (iv) questions to be asked during the group (see Chapter 7 if needed), and (v) the analysis strategies (see Chapters 5, 8, and 10).
2. Each student group should share their plan for critique and make suggested revisions.
3. Select one or two of the plans and role-play the focus group.
4. Use the "participants' responses" and draft a report summarizing the findings.

EXERCISE 12.2   Learning from Practice — How a Needs Assessment Looks

Find a needs assessment report online. Consider using a search engine to search the term *community needs assessment report*. If you have a specific area of interest, include that in the search term (e.g., community needs assessment + English as second language).

Section A: Getting Started

After reading the report, answer the following questions:
a. What was the need being investigated?
b. What type of need?
c. What was the sample?
d. Were all the relevant stakeholders included? Were any excluded? If so, who?
e. Identify the methods used.
f. If the study used either quantitative or qualitative methods, how could it be augmented by the alternate method? For example, if it used qualitative methods, how would using quantitative methods have added value to the findings?
g. What were the major findings?
h. Did the findings as stated offer solutions or remedies for the need?

EXERCISE 12.3   On Your Own

Identify and interview someone in your agency/community who has conducted a needs assessment using the following semi-structured interview guide (feel free to add your own questions and don't forget to use probes).

a. What is your job title?
b. What were the factors that made a needs assessment necessary?
c. Who initiated the needs assessment?
d. Who designed the methodology?
e. What methodology was used to conduct the needs assessment?
f. What were the factors that contributed to using survey, secondary data, focus groups, interview over some other method?
g. What was the most challenging component of the needs assessment?
h. What was done as a result of the findings?
i. If you had to do it all over, what if anything would you do differently?

Working with Geographic Information Systems*

Irvin B. Vann

Geographic information systems (GIS) provide managers and researchers an entirely different way to organize and analyze information. Instead of examining tables and graphs a GIS presents the characteristics of neighborhoods, counties, and other geographic units. This information helps to identify problems that may be unique to a specific location or to target interventions to an area most in need. If you have used Google Earth, you have seen a GIS output. If you have conducted an online search to locate a restaurant chain, you may have been directed to a map with pins to mark each location. GIS produced this pin-filled map.

GIS originated as a tool for studying natural resources, and local governments have adopted GIS as an important management tool. Governments use them to locate utilities and transportation routes, monitor crime, and identify where services are needed. Nonprofits use GIS to target and monitor their fund-raising and to locate, provide, and assess their services. New uses continue to emerge as GIS become more readily accessible to non-technical users, such as community advocates and agency interns. Adaptations of GIS are widely available, allowing for a "do-it-yourself" approach to common tasks that a few years ago required a visit to the local planning department.

The development of GIS software has followed the general trends in computing. Initially, GIS software required extensive training and even special equipment. If you wanted to use a GIS you had to know how to enter instructions line by line into the computer. As the desktop computer became more powerful in terms of memory, speed, and computation, GIS-software makers added more and more user-friendly operations to the software. GIS users today can use the same desktop

---

*Carolyn J. Harper of Wake County, North Carolina, provided examples of local government use of GIS. The authors appreciate her contributions.

computer they use in their other daily activities. Users can enter instructions by clicking commands from menus. They can look up property records and add information selecting from a menu provided by a public-user site. Map information available to public users include topography, building footprints, lot layouts, and street center lines. These tools bridge the divide between technical experts and program staff. A desktop GIS allows you to analyze your data without having to wait for or contract with a GIS technician.

In this chapter we introduce you to basic information on GIS, how a GIS organizes information, and ethical and management issues associated with GIS. A GIS is especially helpful in doing community assessments. As you read this chapter you may see other ways you can use maps to identify problems and propose solutions, assess performance, and present information to the general public and decision makers. GIS can also be a useful research tool for managers.

GIS BASICS

A geographic information system is a computer system for recording, managing, analyzing, and displaying geographic data. *Geographic, or spatial, data* are data about places on the earth. In addition to geographic data, GIS technology uses attribute data. *Attribute data* are nonspatial data providing additional information about the geographic data. Spatial data describe pieces of land or parts of the earth. Attribute data describe what is on, or near, pieces of land, such as restaurants or government buildings.

Two types of geographic data are used in GIS: raster and vector. *Raster data* use individual cells to create an image. This data is used in satellite images, digital photography, and aerial views of land. Scientists and managers use raster data extensively in environmental programs. If they apply for federal funding they may have to do an environmental review and include an aerial view of a site as part of their grant application. Alternatively, *vector data* use points, lines, and polygons to create images. Vector data display streets and roads, and vector polygons display areas with boundaries such as cities, counties, and other jurisdictions. Program managers in the public sector use vector data most often.

Attribute data in a GIS, as mentioned, provide information about a place or location. An attribute map can show where resources are located relative to other resources. For instance, individual maps may show the location of community resources such as schools, licensed foster homes, child care facilities, or group homes. Attribute data can be combined to visually locate places of interest with data on population density, distance from health services and grocery stores, or percentage of low-income households. Community leaders, including agency staff and citizen advocates, may use these data to identify where programs exist and where others are needed. These data are stored along with the geographic data in a database management system (DBMS).

Attribute data must have a link or connector to the land they describe. Spatial identifiers or addresses are necessary to identify places. In the event of a natural disaster or other event, an emergency response dispatcher might ask a 911 caller, "What is the address of the emergency?" The caller would then provide the street address, a common spatial connector, allowing assistance to be dispatched to the area. If the address is 123 Main Street, information on that address could be accessed later as needed. Spatial identifiers like street addresses usually will be linked to other data, such as property ownership, usage, and history associated with them in the GIS record.

Selecting, Creating, and Using GIS Data

As you begin using GIS you will want to learn when and how to get spatial data and decide what data to get. Creating spatial data for GIS is labor intensive and expensive. Therefore, acquiring spatial data may be difficult for nonprofit organizations and other nongovernmental bodies. Fortunately, organizations such as the U.S. Bureau of the Census and state geodata centers make their products available for free over the Internet or by compact disc (CD). Other sources of local area spatial data are local government planning or tax departments, which may charge only a nominal fee for data. Spatial data created by local GIS departments are generally more detailed and more accurate than the spatial data available through national or state agencies.

GIS data are useful in identifying property information. Local land use and zoning ordinances require that changes to property be recorded with the register of deeds. This requirement provides an official and historical record of the literal lay of the land as map records and deed information are connected to property records. This information is affordable and valuable if you are charged with finding a building site or investigating the ownership of a piece of property.

One of your initial choices in using a GIS is to decide what area you want to examine. Data displays for a large area blur specific details while too much specificity may cause you to miss important characteristics. For example, by looking at data within only one small area, you may not see similar activity in a neighboring area. Social conditions may not stop at an artificial boundary like a neighborhood, city, or county limit.

Basic GIS Analysis

The two types of data stored in a GIS, attribute and spatial, provide opportunities for a variety of analyses. Both data types are stored in a relational database. They can be used separately or together to look for relationships. The relational aspect of the database management system allows the user to "query" the attribute data to look for new relationships. To describe clients receiving certain medical services, for example, you could query the system for the number receiving services by age group or income and include a geographic location such as the neighborhood. The GIS display creates maps showing the variation in the number receiving a service by age or income and location. A GIS display is useful as communities seek to identify where programs exist and where they are needed. A needs assessment to identify services for homeless children might include a map showing the location of existing foster homes.

Spatial data are analyzed differently than attribute data. Managers will find four types of spatial analysis most useful for vector data: buffering, intersection, union, and dissolving. Each can be used alone or in combination to explore the geographic impact of a program. We describe these functions briefly in the following paragraphs.

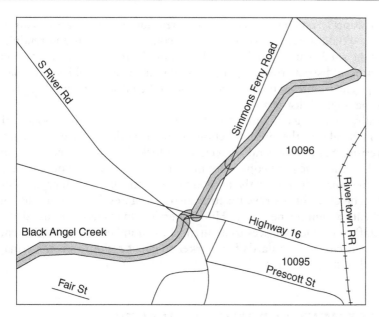

FIGURE 13.1

Buffering Along Black Angel Creek, Riverton

Buffering in GIS is the capability to define the area around a point, line, or a polygon shape. Buffering also separates land uses and sets limits around a building such as a school, house of worship, or child care center. One example of buffers is riparian buffers established around streams to protect the watershed. For example the fictional town of Riverton has established a 50-foot buffer around Black Angel Creek, which feeds into Biggs River. This buffer is shown in Figure 13.1 as the shaded area along Black Angel Creek.

Decision makers may use buffering to show prohibited land uses. For example, many states have laws delineating public buildings as "smoke free" (e.g., Maryland, Delaware, North Carolina). These ordinances require a perimeter or buffer zone of 25 feet from the entrance, within which individuals must refrain from smoking. GIS technology can easily overlay a buffer or shaded area around public buildings to determine the appropriate place to post "no smoking signs" and provide ash and trash receptacles.

The *intersection* function finds commonalities between two files and their attributes. Intersecting creates a new file using only the attributes and features that the files have in common. For instance, intergenerational supportive services are becoming an issue in many areas. Grandparents may become primary care givers in families affected by HIV/AIDS, incarceration, substance abuse, and other social problems. Both the grandparents and children may need social services. A program manager could use the intersection function to identify overlapping child care and senior service centers and identify opportunities to move centers closer together.

The *union* function combines the shapes of two or more areas into a new area that has the boundaries and characteristics of the original areas. Suppose a program manager wants to combine several service areas into one large area to improve efficiency. He would use the union function to combine the areas and retain all of the data for the smaller individual areas. For example, if the union function created a unified school district from two districts, the manager would see the data on all school age children from both districts. The unified district would keep the boundaries of the original districts.

Finally, *dissolving* combines smaller areas and removes their boundaries. This function is similar to the union function but requires that the areas have a common characteristic. Suppose several adjacent city blocks include a concentration of special needs clients, such as people with mobility issues. With the dissolving function a program manager can view the total area the clients live in rather than looking at individual blocks. The dissolve function also aggregates the total number of special needs clients within the new area. Human services managers may use dissolving to assign staff to provide services to an area. If the number of potential clients in the area exceeds current standards for worker to client ratios, a manager may assign additional staff to that area.

MANAGEMENT AND ETHICAL ISSUES ASSOCIATED WITH GIS

General Issues about Maps

Desktop GIS have the ability to create maps easily to illustrate relationships; however, you should understand the shortcomings of maps. All maps represent their author's point of view. Ultimately, any GIS used to analyze and display program results will have some bias, unintentional or otherwise. Monmonier[1] identifies four general aspects of mapping that influence the interpretation of maps. The four aspects are scale, projection, symbols, and generalization.

Map scales can be confusing with regard to the difference between a large scale and small scale map. Generally, map scales are expressed as ratios. Numbers like 1/5,000 and 1/50,000 are examples. Translated to mapping, ratios are read respectively as 1 map inch equals 5,000 inches on the ground or 1 map inch equals 50,000 inches on the ground (4,166 feet). Choosing the right scale is important to avoid distortions; more detail can be shown when the denominator is smaller. A ratio of 1 map inch equals 100 inches on the ground provides a more realistic picture than does a ratio of 1 map inch equals 5,000 inches on the ground.

Maps are imperfect because they are two-dimensional representations of the three-dimensional spherical earth. The transition from earth to map is called a *projection*. Mapping uses many types of projections, which all seek in some way to create a balance between preserving shape and distance. Projection becomes more important when you overlay data from different sources. For example, if a program manager overlays a transportation map with a map of recreation sites, the

---

[1]Monmonier, Mark; Ethics and map design: Six strategies for confronting the traditional one-map solution, *Cartographic Perspectives* (1991), 10:3–8.

projections must be the same. Otherwise, the position of the recreation sites will not match where they are located on the ground and will be inconsistent with the transportation map.

Symbols directly affect the interpretation of maps. They represent the data being displayed and are influential in communicating messages to others. Symbols, for example, buildings, steeples, and bells, might be used to show the locations or numbers of government offices, churches, and schools respectively. Symbols also provide spatial orientation for looking at maps. Generally, maps are oriented with the north arrow positioned at the top of the map. Whether or not a map shows a north arrow, you can assume that a map is oriented with the north arrow up. Program managers who are beginning to learn a GIS should use the keep-it-simple approach when using symbols. This is especially true if they are presenting the maps to clients or stakeholders. No audience wants to be overwhelmed with symbols that may be difficult to interpret.

Finally, there is the issue of *generalization*. Generalizing map data means to smooth or leave out certain levels of detail: too much detail and the map may be difficult to read and therefore not useful. However, if you include too little detail, important characteristics may be omitted. As with any other presentation, generalizing is a choice you will need to make based on what you think is important.

Privacy Issues in GIS

One issue that concerns program managers is privacy. If services are delivered to clients at their home addresses, showing where services are delivered on a map may violate clients' privacy. In general, the more that location data are aggregated, the less likely the chance of violating someone's privacy. However, aggregating data does not always guarantee privacy. In an area with few cases, others may be able to disaggregate the data sufficiently to identify specific individuals, who then are no longer anonymous. If this happens, participants may be reluctant to share any data in the future.

Common GIS Mapping Functions

The ability to create and use maps is the core strength of a GIS. Program managers have a wide array of possibilities regarding how they apply the mapping capabilities of GIS. The following eight mapping functions commonly used in GIS analysis are available in desktop GIS software and require a minimum of technical training. Their results are usually easily understood by a wide range of users, not just by managers or technical specialists. The functions are pin mapping, hot spot mapping, density mapping, presentation or briefing maps, pattern detection, proximity mapping, interagency maps, and decision maps. These functions provide immediate feedback about program variables in relation to their geographic areas. Additionally, these functions are useful for creating maps that convey the spatial relationship between program outcomes and selected independent variables. In the context of research, managers can use information about spatial attributes as independent variables to investigate relationships to program outcomes as dependent variables.

Pin Mapping: Pin mapping is the electronic version of sticking pins into a map and looking for visual patterns in relation to their geographic locations. The software finds an address in a street database and places a symbol on it. It is a basic function found in every GIS package. Multiple activities shown at the same address or close to an address, however, can create visual clutter on maps. For example, mapping multiple activities or programs at the same address can create confusion if the mapping scale is too small and the "dots" appear to merge. If the dots are the same color, the map user may not be able to differentiate between them. GIS can solve this problem, since most software packages have the ability to change the shape, size, and color of the dots.

Pin mapping has the highest potential to generate privacy concerns. Consider the loss of privacy if each pin represents the address of a child who was placed in foster care. The dots or pins highlight a smaller area of land, making it easier to guess an identity or exact address. Sometimes the dots are offset or placed in the center of a road to disguise the exact location. Note the dots in Figure 13.2. They represent an event, activity, or person that can be identified by a specific address such as a street address.

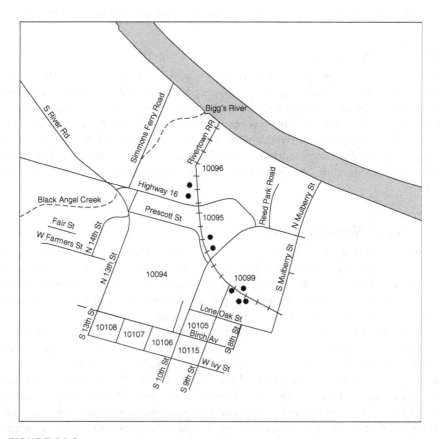

FIGURE 13.2
Riverton Pin Map

Hot Spot Mapping: Hot spots are incidents or points of activity that have been smoothed to create density areas. The points could represent crimes, incidents of disease, fatal automobile accidents, or any other activity that has a specific address. After calculation, the density areas are shown ranging from high to low density. Hot spots could be the areas where program services are most required or where a program will make its greatest impact. For instance, Child Protective Services (CPS) workers might find that a particular neighborhood is a hot spot for CPS reports. Hot spots signal higher density and a higher probability that an incident of interest will be found; hot spots can also represent areas where a program is having its greatest effect.

Hot spot mapping has been used to investigate the distribution of services like prenatal clinics. For example, researchers in New York City found that women using prenatal clinics in Brooklyn were more densely distributed in the north and central parts of the borough. Access to public transportation was a critical factor in going to a prenatal clinic. Consequently, administrators considered two options: construction of prenatal clinics in the north and central parts of Brooklyn or increasing transportation access in those sections

Hot spot mapping is used extensively by public safety agencies. Police agencies use hot spot mapping to focus on areas with the greatest possibility of criminal activity and to adjust their policing strategies according to the prevalent types of crime. The city of Baltimore has used hot spot mapping to determine where to institute community policing as opposed to surveillance cameras. Local police, child welfare agencies, and juvenile court counselors can map hot spots where juvenile arrests occur and develop a coordinated response.

Density Mapping: Density mapping is used for mapping activity that is continuous in a geographic area. The unit of analysis is an area; a program manager can use density mapping to determine which county areas have the greatest concentrations of low-income families or teenagers. With this information, a program manager could establish services based on a certain density threshold in an area, for example, providing services where there is greater than 20 percent of a certain demographic. Your ability to use density mapping is limited by the level of data collection. The data may not have been collected in a small enough area to be helpful. Figure 13.3 illustrates Riverton's density map. The shaded areas show where there are high concentrations of abandoned buildings and buildings that do not meet building codes.

Another issue is the stability of the boundaries. Accurate data at neighborhood levels are sometimes difficult to obtain. One of the first suggestions a manager may hear is to use U.S. Postal Service ZIP codes. However, ZIP code boundaries are not stable; they are changed to fit the business needs of the postal service. A ZIP code boundary could incorporate a certain area in one year; and in the next year parts of the same area may be in another ZIP code.

When data from areas with stable boundaries are required, analysts use data from the Bureau of the Census. The boundaries stay the same from decade to decade, and the data are free. Census blocks are drawn by the Census Bureau at a smaller, neighborhood level. Census blocks are aggregated to create census tracts. However, this aggregation can result in some misleading conclusions since the tracts may contain blocks with outlier values. For example, in a mixed zoning area,

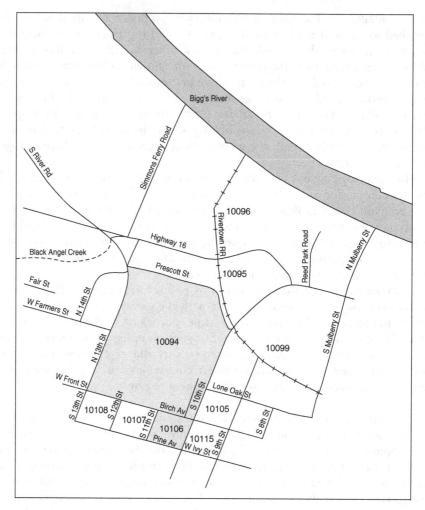

FIGURE 13.3
Riverton Density Map

a high-income block or a low-income block might skew the income values of the tract as a whole. However, many researchers use census blocks as a proxy for the concept of neighborhood. While the boundaries are stable, the data collected are not. The questions asked by the Census Bureau vary from census to census.

The differences between ZIP code data and Census Bureau data have created a market for a product called the "ZIP to Tract" conversion. These products match ZIP codes to census tracts. Additionally, these commercial products may provide data collected at the ZIP code level that the Census Bureau does not collect.

Decision Maps: Decision maps are useful for displaying selected aspects of a program within a geographic area. Program analysts and managers may use these maps to determine which aspects of a program are successful in particular areas.

As noted before, the limiting factor is the availability of the data by a geographic boundary. For example, with current GIS capability on desktop computers epidemiologists can study the social factors contributing to the risk of sexually transmitted diseases among young adults. By finding a model that forecasts which areas have the greatest risk, prevention efforts can be focused where the need is greatest. Or by using data on the harmful effects of industrial waste products, researchers can project the number of cases of respiratory diseases that will occur in residential developments built on or near former industrial plants. These methods do not predict events with perfect certainty. But they do provide information guiding decisions as to where to place services and to what extent to monitor producers of potentially harmful substances. In many ways, this type of analysis is very similar to weather forecasting.

Decision maps can model the impact of making a decision. For example, more than one method may be available to achieve a program outcome. To help make that choice, you can use GIS to model different choices and their anticipated outcomes. For example, you could compare different sites for an industrial plant to estimate their environmental impact and the cost of mitigation. Based on your findings you can choose the method that does the least harm or has the most favorable benefit-to-cost ratio.

Presentation or Briefing Maps: We are bombarded daily with various briefing maps. Many communities publish their crime statistics on the Web using maps so that citizens are aware of crime reduction efforts in their neighborhoods. Television news programs show maps to illustrate daily events and trends. The oldest briefing map used on a daily basis is probably the weather map. If you live in a large city, you may see a briefing map when you watch the morning traffic report.

Although similar to decision maps, presentation or briefing maps are messages communicated to superiors, clients, or to the public at large. Usually these maps have a persuasive point or message the authors want to highlight. A good example is when properties are rezoned. Usually, a planning department holds public meetings to present a rezoning proposal to affected property owners. The briefing maps give property owners an opportunity to see and comment on the effect of rezoning before a city council makes a final decision on the rezoning request.

Presentation or briefing maps transform data into information. They summarize an otherwise large amount of data into something more understandable and visual. For example, policy makers deal with immense amounts of information and numerous requests for funding. Program staff may have only a limited opportunity to request funding for their agencies or projects. Briefing maps streamline communications from staff, with the added benefit to managers and other policy makers of having visual representations of the information

Pattern Detection: Pattern detection looks for similar events and circumstances within a geographic area. Maps may identify the location of conditions indicating a need for services. For example, a map may show areas with a concentration of CPS cases, which in turn may suggest where to locate family support services. Simple pattern detection is often used with pin mapping or density mapping. This usually involves looking for dots that are clustered closely and distinctly in a geographic area.

As researchers have become more interested in events and their geographic locations, various pattern detection methods have been developed to determine whether events, such as fatal teenage auto accidents, are random or geographically based.

The general rule in pattern detection is known as Tobler's first law of geography,[2] named for Dr. Waldo Tobler, who pioneered this concept in his research. Tobler's law is usually stated as "everything is related to everything else, but near things are more related than distant things." Tobler's law seems to hold up in situations when local conditions are closely correlated with the events. Events like disease and crime often are rooted in local conditions. Spatial statistics, a branch of statistics for geography, are used to determine whether these patterns have some special significance in relation to their locations.

Proximity Mapping: Proximity mapping illustrates how far something is in relation to something else. Proximity maps are good for answering questions like "how close is a program to the customers?" or "Is distance from certain programs an important variable to their effectiveness?" For example, how far is an afternoon tutoring program from a middle school or high school? How far do retirees have to travel for medical services? Although the hot spot mapping technique can answer some of these questions, proximity mapping methods are more intuitive and easier to present to a general audience. One proximity method uses concentric rings with radii centered on the object of interest. You may decide to look for a program's potential clients within two miles of its location in half-mile increments. Concentric rings are useful for examining events that may extend in all directions from a location. In the case of populations living near a nuclear power plant, for example, safety or evacuation zones are usually shown on maps as series of concentric rings radiating out from the facility.

Proximity mapping can be used to plan for capital improvements and service expansions. Consider a human services center with five locations where citizens can receive Medicaid and other benefits. Suppose that the data indicated that citizens from a nine-square-mile sector of the county did not seek services from any one location. When the organization decided to expand, proximity mapping established the case for creating a mini-center to serve that community.

Most GIS software programs have the capability to measure linear distance from one point to another. This is the "as the crow flies" straight-line distance as opposed to the road or "on the ground" distance. Remember the straight-line distance will always be less than the road distance. If you choose to use this type of map, you will want to make it clear how you determined the distances.

Network analysis is another proximity method. Instead of measuring distance in a straight line independent of roads, network analysis measures distance along road networks. Network analysis can determine estimated travel time and distance between points in the road system. One familiar application of network analysis is the portable GPS/GIS found in some automobiles. When a GPS/GIS determines the shortest distance or time to your destination, it is using network analysis.

---

[2]Daniel Z. Sui, Forum: Tobler's First Law of Geography: Big Idea of a Small World, *Annals of the American Association of Geographers*, (2004), 94(2):269–277.

Program managers may use network analysis to manage service delivery. For example, towns and cities use network analysis to balance the workload for the sanitation department. School bus managers use network analysis to make bus routes more efficient. Depending on the GIS, program managers may choose to optimize routes based on either time of day or traffic patterns. As in other GIS functions, the analysis depends on the accuracy of the data. When using network analysis, the best or optimized route should be verified for accuracy. For example, you do not want to schedule an activity that is going the wrong way on a one-way street.

Interagency Maps: An important task in public sector work is coordinating information from several agencies. You can display data from more than one agency on one map. For example, public transportation data can be overlaid on data from the recreation department. The recreation department can then determine the best location for a recreation center so that it is accessible by public transportation.

An interagency map can serve two purposes. Initially, a program manager can use the map to highlight how a problem crosses departmental or agency boundaries. An interagency map can help identify where separate agencies have overlapping interests. With this information, potential partnerships may form to undertake a project or to join together after a project is completed and running on its own. Later an interagency map can highlight the success of the partnership and how each partner contributed to the project's success.

A key technical issue with interagency maps is making sure the data have the same scale and projection. Otherwise, the spatial relationships will be distorted. If the partner agencies are using data from the same source, for example a shared mapping department, then scale and projection should not be a problem.

USING GIS FUNCTIONS: EXAMPLES OF APPLICATIONS

GIS functions allow program managers to create maps to view important social and economic characteristics of their service areas and to combine the maps in different ways. The analytical and mapping functions also allow managers to investigate the relationship of selected program variables to the spatial characteristics important to program activities.

Managers can use GIS information along with knowledge of program goals and objectives to plan, manage, and evaluate programs. For example, a community assessment team recommended that Child Protective Services (CPS) strive to better coordinate services between schools and foster homes. GIS staff members first used program data to create an attribute map of all licensed foster homes in the county. Next, they created a map showing interstates, state highways, and secondary roads and a pin map of the locations of public schools; the two maps were overlaid. The staff used the GIS union function to group adjacent, contiguous, and abutting neighborhoods. Working with maps similar to those shown in Figure 13.4, the CPS could explore opportunities for partnerships between foster homes and a school. CPS staff could target its efforts to increase foster homes in areas with schools and few foster homes.

Analysis of the program data showed that some foster homes had empty beds, and some neighborhoods had very high numbers of children who were not placed in nearby foster homes. The analysts found that twice as many children were in

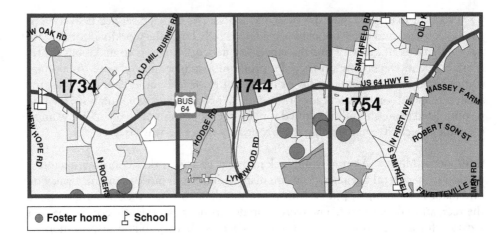

● **Foster home**   ⚑ **School**

FIGURE 13.4
Location of Foster Homes and Public Schools

foster care as the number of licensed foster home beds in the county. Program managers used GIS maps to guide efforts to increase the number of licensed foster homes within the county and place them in areas with high need for beds.

Using statistical software in combination with the mapping capability of GIS has emerged as an important trend in creating decision maps. Analysts use statistical processes to create models to forecast events based on a given set of conditions including geography. Remembering concepts from Chapter 1 you should be able to label attributes as independent and dependent variables and think about how they are related to each other. GIS helps you to identify and investigate the special attributes of areas. For example, in a crime analysis study in Minneapolis, Minnesota, researchers noted that 50 percent of the crime occurred at 3 percent of the city addresses. The study determined that certain classes of criminal activity and certain addresses were connected.

Another study in Maine mapped vehicle crashes involving alcohol and residents under age 21. In both examples we see that certain variables were related to specific locations and may be related to other attributes of those locations. You can use GIS and statistical routines to calculate traditional measures of association between geographic attributes. For example, the rate or type of crime may be correlated to the average age of people living in an area. Vehicle crashes involving alcohol may be correlated to the location of establishments rumored to sell to underage customers. The location of bars and restaurants selling alcohol may also be related to the location of a particularly unsafe road.

CONCLUDING OBSERVATIONS

The purpose of this chapter was not to make you a GIS expert. It was to introduce you to some basic concepts about using GIS for program management. The concepts described in this chapter were selected because most GIS software programs are capable of creating pin maps and density maps and determining hot spots. With these tools a program manager can create briefing maps, decision maps, proximity maps, and interagency maps. These maps will be familiar to program managers as well as stakeholders since they appear in the televised and printed media to illustrate news stories. The familiar look of the maps promotes a level of comfort in using GIS as a tool and creating products using it.

Managers can find GIS to be an important addition to the research methods discussed in this book. Applying GIS as a management and research tool often involves identifying a dependent variable of interest along with independent variables influencing the dependent variable. The reason for using a GIS is that one or more of the independent variables has a spatial or geographic relationship to the dependent variable. Desktop GIS software gives program managers tools to examine the impact of the spatial variables on dependent variables of interest. The dependent variable could be any measurable program output. Patterns suggesting relationships among variables that might not

otherwise be identified may become obvious when viewed on a GIS-produced map. Additionally, GIS allows a program manager to communicate using maps. Often, clients and other stakeholders find it easier to understand a map rather than tables, charts, or graphs. For some stakeholders, especially politicians, seeing a problem or a solution on a map may motivate action.

GIS analysis, however, still requires program managers to understand their programs and the variables affecting them. The old computer adage of garbage in, garbage out (GIGO) applies to GIS analysis as well. A carefully constructed map with clear analysis communicates impact effectively. However, poorly constructed analysis—even if the map is a work of art—confuses and ultimately affects the credibility of a program. By carefully identifying program variables, choosing valid measures, and carefully collecting data, managers can avoid the dangers of GIGO and use GIS to expand their ability to obtain useful results.

RECOMMENDED RESOURCES

Peters, Alan, and Heather MacDonald, *Unlocking the Census with GIS* (Redlands, CA: ESRI Press, 2004).

Price, Maribeth, *Mastering ArcGIS*, Fourth Edition (New York: McGraw-Hill Companies, 2010).

Sommer, Shelly, and Tasha Wade, *A to Z GIS: An Illustrated Dictionary of Geographic Information Systems* (Redlands, CA: ESRI Press, 2006).

Tomlinson, Roger, *Thinking About GIS: Geographic Information System Planning for Managers*, Third Edition (Redlands, CA: ESRI Press, 2007).

At the bottom of the Web site http://www.abcdinstitute .org/resources/ is a section called Mapping Tools, which provides downloadable worksheets.

CHAPTER 13   EXERCISES

The four exercises in this chapter will develop your ability to identify opportunities to use GIS to make program decisions.

- Exercise 13.1 Creating an After-School Program asks you to link mapping functions to identify and describe possible program locations.
- Exercise 13.2 Services for the Mobility Impaired asks you to describe how you could locate impaired residents and their access to services.
- Exercise 13.3 Learning from Practice: How Agencies Use GIS asks you to identify a report that uses GIS to see how and why GIS was used.
- Exercise 13.4 NewLand asks you to plan the GIS components for an urban development project

EXERCISE 13.1   Creating an After-School Program

Scenario

Your agency is preparing a proposal to establish after-school day care for children and teenagers from ages 6 to 16. As a condition for establishing the centers, parents have petitioned the city council to incorporate the following conditions:

- All centers must be located at least 100 feet from any four-lane city street.
- Centers for children aged 6–12 years old must be no more than a 15-minute ride by city bus from their school to the center.
- Centers for teenagers aged 13–16 must be no more than a 15-minute ride by city bus or a 25-minute walk from their school to the center.
- No center will have any registered sex offenders living within a 1,000-foot buffer zone.

Section A: Getting Started

1. Identify one or two of the mapping functions typically included in a desktop GIS that you could use to address each condition.
2. What other agencies you would want to contact (a) to obtain information or (b) to evaluate your proposal? Briefly describe the reasons for contacting these agencies.
3. Describe the GIS map you would prepare to present at public meetings.

Section B: Small Group and Class Discussion

1. In groups of three to five review and discuss each member's proposal. Then develop an outline for the proposal that you will present at public meetings.
2. The entire class should review each group's proposal. Based on its discussion the class should develop guidelines entitled Considerations in Creating Effective GIS Maps and Presentations.

EXERCISE 13.2   Services for the Mobility Impaired

Scenario

A Department of Social Services (DSS) has the addresses of mobility-impaired clients who live in the county. A staff member wants to develop a system of regularly scheduled transportation to help these clients shop, work, and have access to other services.

Section A: On Your Own

1. Describe how you would use GIS to
   a. identify the addresses of sites where the clients can shop, work, and have access to a neighborhood DSS office.

   b. plan a route for a shuttle bus to circulate through the area where the clients are living.
   c. minimize the travel time from clients' homes to shopping, working, and other services
2. To create maps for the staff consider how and why you would use the intersection, union, and dissolving functions.
3. What other agencies are you likely to coordinate with in order to complete the project?
4. Describe the format you will use to visually present this project at public meetings, such as a meeting of county commissioners.

Section B: Class Exercise

The entire class should review the proposed formats and discuss if it needs to make changes to Considerations in Creating Effective GIS Maps and Presentations.

EXERCISE 13.3   Learning from Practice: How Agencies Use GIS

Go to a search engine and find a report of an organization's use of GIS. Try a search term such as "Examples of GIS Projects." Or to narrow your search, try "Nonprofits and GIS" or "GIS in Local Government."

Section A: Getting Started

Prepare a summary of the report to share with your classmates. Include the following information about the project you are reporting. Note in your summary if the report did not contain the information requested.
a. The nature of the organization using the GIS.
b. The nature and purpose of the project reported.
c. How many organizations were involved and their types (public, non-profit, for-profit). Also note where the organization obtained the data for the project. Did they collect it or did another organization collect it?
d. The name of the GIS software used.
e. Were maps included in the report? How would you identify them according to the types of maps discussed in the chapter?
f. If a map was included, what was the map scale?

Section B: Small Group Exercise

1. In groups of three compare each member's report and note the following:
   a. types of GIS software used
   b. data sources
   c. types of GIS maps included in the report
2. As a class, compare the findings of the groups and discuss if their findings suggest trends in popular software, data sources, and types of presentations. What else did class members note that may be valuable in applying GIS?
3. Each small group should select one member's summary and locate additional information about the area covered in the reported project, for example, the area's population size, degree of urbanization, and

unemployment rate. Each group should also locate a printed or Internet map of the area covered in the report. Based on the information found, answer the following items.

a. What, if any, of the features on the printed or Internet map was also included on the map in the report or mentioned in the report?

b. Compare the map scale of the printed map with the map scale in the project report.

c. Suggest mapping functions that were not used in the report but that you think could have been useful.

EXERCISE 13.4   NewLand

Scenario

NewLand is a nonprofit organization working to help business, government, and other nonprofits find better uses for land in Town Place, an old suburb in a metropolitan area. NewLand would like to see new residential development in Town Place with nearby access to shopping and jobs. It also plans to rent a facility where it will provide job training for local residents. The organization purchased GIS software and had two staff members trained in GIS, but data were never collected because of limited financial support and staff time. Your professor has contacted NewLand and offered the class's help in data collection.

Section A: Getting Started

a. List variables that NewLand should collect data on (i) for its map of Town Place's neighborhoods and (ii) to track whether NewLand is accomplishing its objectives.

b. Suggest what images in the form of maps, charts, and pictures would be useful to planners who are developing the area. Identify at least two types of maps discussed in the chapter.

Section B: Small Group and Class Exercise

1. Working in groups of three to five class members,

a. review the individual lists of variables and combine them into a single list of variables. Explain how each data item will serve to meet NewLand's objectives.

b. Identify the information you would present to planners in Town Place and recommend a format.

2. The class should review Considerations in Creating Effective GIS Maps and Presentations and consider if it should be redrafted or if a separate set of guidelines should be developed.

Communicating Findings

S trategic managers and entrepreneurial researchers know that unless research findings are reported effectively nothing happens. You may report findings at meetings, in press releases, brochures, project reports, annual reports, and academic papers. In this chapter we argue for clear, focused presentations tailored to the needs of a particular audience. At the end of the chapter we touch on ethical concerns that occur in connection with reporting data.

VARIATIONS IN AUDIENCES AND THEIR NEEDS

Whether you plan to make an oral presentation or write a report, the first steps are to focus on your purpose and the characteristics of the intended audience(s). First, you need to get the audience's attention. Virtually everyone is overloaded with information. Effective administrators and policy makers may be particularly adept at protecting their time and ignoring information that they do not need or want. Second, you want to prevent having the listeners or readers miss your main point. If you aren't clear and don't get their attention they can misunderstand or ignore important findings. Third, you may want to teach your audience members something, influence their thinking, or motivate them to act.

Identifying the audience for an oral presentation is straightforward. Your first question should be "why will people attend the presentation?" Once you have the answer you can tailor the presentation to the audience's concerns, its level of knowledge, and its motivation to act. Identifying potential readers is more difficult. A report may be passed on to supervisors, staff, agency analysts, interest-group members, professional acquaintances, legislators, or students. Reports may be placed in an agency library or posted on a Web site. To satisfy diverse readers, reports must be clearly written and research procedures should be fully documented. On the other hand, including full details, especially about the methodology, can diminish a report's readability or an audience's attention. You can resolve this apparent conflict by putting important information, such as the report's findings and recommendations, first and placing complicated or technical details in footnotes or appendices. You may also direct an audience to Web sites or other easily accessed sources for additional information.

If you are conducting a study for an organization that you don't work for, you may want to learn how it normally organizes and presents information. You may attend oral briefings or ask potential audience members to identify presentations that they thought were especially effective. You can use this information to infer what features generate audience interest and involvement. When you read reports, save ones that seem particularly well done.

ORAL PRESENTATIONS OF RESEARCH FINDINGS

No matter whether you see yourself as a strategic manager or an analyst, you need to hone your oral presentation skills. As teachers we have observed talented students who avoid making oral presentations. These students lose valuable opportunities to practice presenting their ideas, listening to others, and phrasing and answering questions. Whether you normally speak to one person, a small group, or a large, formal audience, your ability to explain your work clearly will serve you well. Professionals who feel pressed for time may prefer to hear about a study rather than read through a report. Some people are "oral learners," that is, they efficiently absorb and understand information they hear. Others value debating information and discussing it with investigators and colleagues.

An oral presentation provides an excellent opportunity to have an impact. Audience members may feel compelled to pay attention. Their interactions may motivate the group or individuals to discuss and follow up on the findings. You can prepare by asking yourself, "Why will people attend this presentation? What do I want them to learn? What action do I want them to take? How can I convince them to take it?" The answers should guide how you organize a presentation.

You should not discount the importance of one-on-one informal discussions of your research. Their informality can be deceptive—don't overlook the opportunities they provide. They offer an important occasion for others to develop interest in your project. What would you say if your agency head were to ask what you are working on? Wouldn't you want to generate interest in your current project? Wouldn't you want to lay the groundwork for a decision based on your findings? A trick that some researchers employ is to prepare an "elevator speech." That is, they prepare a very short description of what they are working on that is short enough to be said during an elevator ride. You may never be caught in an elevator with a person you want to impress, but if you are prepared you will not waste an unexpected opportunity to sell your project.

An effective presentation requires planning and practice. Select the points you want to emphasize, the evidence you will use to support these points, the order in which the information will be presented, and visual aids. The traditional order for a research presentation—background, methodology, findings, and discussion— usually works well. It develops the material logically. People with training in the sciences, including the social and behavioral sciences, have come to expect it. If audience members are already informed about the program or policy, identifying the study's purpose may be sufficient. Otherwise, you should describe the program or policy to put the information in context and to help audience members follow the presentation. Usually a brief discussion of the methodology is sufficient. Except for specialized audiences, you can skip the technical details. Remember that while you have learned to pay careful attention to detail and to examine findings from various perspectives, these skills can translate into tedious, unfocused presentations. Avoid trying to cover too much information. Instead concentrate on a few important points and encourage the audience to ask about the details, especially those details that may affect their willingness to accept the findings.

Visual aids may be used throughout a presentation. PowerPoint slides, tables, or graphs focus the presenter and the audience. To select a visual aid, consider whether it communicates the information clearly and effectively, requires special equipment, or slows down the presentation. Too many visuals can bore an audience. Detailed tables and graphics leave people in the back rows squinting or feeling left out. Wordy slides focus the audience on trying to decipher the slide instead of listening to you.

The slide shown in Figure 14.1 is from a presentation on the legislative history of a U.S. health care policy. You might rightly point out the slide is too wordy, but even more confusing is that it lacks coherence. We might start our revision by focusing only on the Social Security Act and Wagner-Murray-Dingell Bill. We would give the year of the Wagner-Murray-Dingell bill. The citations probably could be deleted from the slide. Figure 14.2 shows an improved version. Using PowerPoint you might try your hand at further improving the slide in Figure 14.1.

If your presentation contains a number of tables or slides, or if you expect the audience to take notes from the PowerPoints, you should prepare a handout containing the same information. Alternately you may post the slides on a Web site or e-mail them to participants.

Presentations with lively graphics can be fun to put together, but make sure that they don't draw attention away from the presentation's content. Similarly, fumbling around with unfamiliar equipment creates a serious distraction.

Inexperienced presenters may overlook the importance of practice. A researcher who has poured over a study may feel confident in her ability to ad lib the

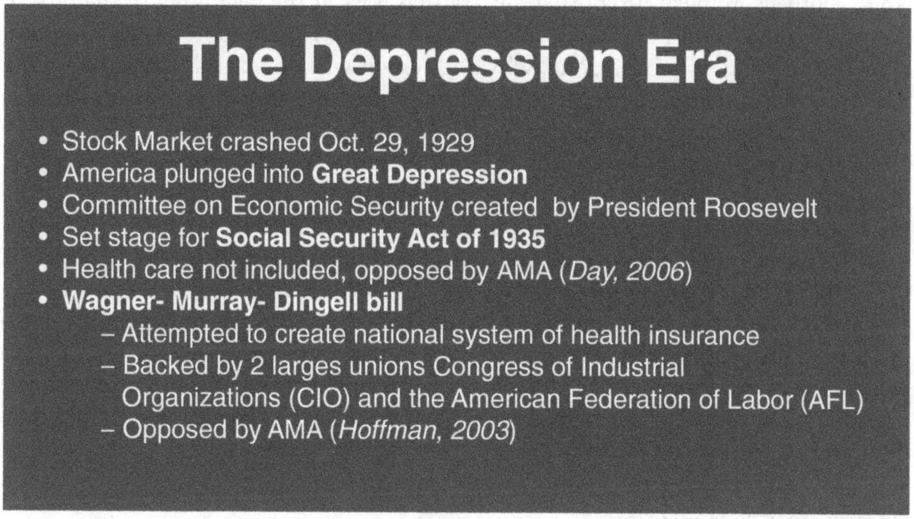

FIGURE 14.1
An Ineffective Slide

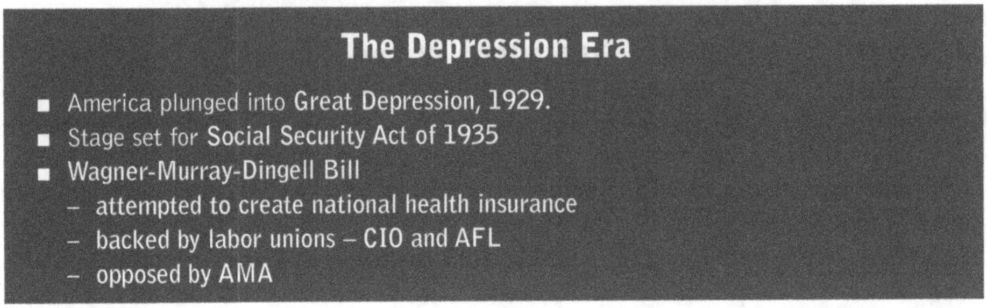

The Depression Era

- America plunged into Great Depression, 1929.
- Stage set for Social Security Act of 1935
- Wagner-Murray-Dingell Bill
  - attempted to create national health insurance
  - backed by labor unions — CIO and AFL
  - opposed by AMA

FIGURE 14.2
A More Effective Slide.

presentation. Unfortunately, she may bog down on the study's minutiae or move erratically from point to point. Typically, one should practice with an audience of colleagues, team members, or friends. Practice-session observers should make sure that the major points are clearly presented, the statement of key points does not become repetitious or condescending, the transitions are smooth, and the equipment operates correctly. The observers should ask questions about the methodology and the interpretation of the findings. Preparing answers to "hard" questions avoids the embarrassment of stumbling around during the actual presentation. If questions challenging the credibility of the study or its findings go unanswered or are poorly answered, the written report may never be read and its potential impact may be undermined.

WRITTEN PRESENTATIONS

A written research report should cover the study's purpose, relevant background articles and reports, its methodology, findings, discussion, and recommendations. An executive summary, a report summary that goes at the beginning of the report, is actually the last part of the report you write. The written report should be a permanent record of what was done, why and how it was done, and what was found. Although the number of people who actually read the report may be small, this written record remains and may be available to all interested parties.

Research findings may be summarized on handouts and Web sites, in brochures or press releases. Typically, to prepare such summaries you extract material from the research report. Readers of summaries miss the details they need to judge the credibility of the findings, to pick up information that may be pertinent to them, or to justify making a decision or taking action. However, the summaries may be the only part of the report that some policy makers and others read. The *executive summary* provides an overview of the important aspects of the research report. It often is included as the first part of the research report and can also be used as a separate document. We first discuss the structure and content of the research report and then discuss the executive summary.

Background Information

You should begin the report by identifying the question you are asking and the value of answering it. As appropriate, a report may discuss the program or policy's origins, implementation history, goals, relevant stakeholders, resources, and activities. The specific information included depends on the report's audience and its purpose.

To develop the background information, you may cite interviews, documents, and the research literature. You are most likely to include a formal literature review in program evaluations. The literature may justify the study's design, the variables you chose, the relationships you examined, and how you interpreted the findings. You may present previous research in chronological order or you may organize the discussion around key variables or concepts. You may weave information from the literature into the background presentation, assign it to an appendix, or include it in an annotated bibliography.

Methodology Section

The final project report should be comprehensive enough for others to use the report, verify its findings, or replicate the research. The methodology section is key to providing this information. At a minimum it should have enough detail so that readers can decide if the findings are credible and policy makers can use them as evidence. You should discuss how you defined and measured the study's variables, any intervention you introduced, your sample, when you collected the data, and how often you collected them.

If your study design was an experiment or a quasi-experiment you should describe the intervention, the study population, and how you created the study group(s). Your goal is to provide sufficient information on the design and its implementation so readers can assess the study's internal validity, and subsequent investigators and policy makers can assess how the findings might apply in other settings. When reporting performance measures or survey results you can limit the methodological discussion to writing about the measures and samples.

In the measurement section you should identify the operational definitions, how you categorized or assigned numerical values, how you grouped values and combined variables to create indicators, and evidence supporting the reliability and operational validity of the measures. Customarily, in quantitative studies researchers report only the findings from mathematical tests of reliability and empirical evidence of operational validity. To illustrate what is included we give a hypothetical example of how to report the operational definition. If an analyst

---

▶ REPORTING ON A MEASURE

To measure trust we asked respondents to rate the following statements using a 7-point scale where $1 =$ strongly disagree and $7 =$ strongly agree; the alpha coefficient was .88.

- The people who represent the funder are trustworthy.
- My organization can count on the funder to meet its obligations to the program.
- My organization feels it worthwhile to continue to work with the funder. ■

divided the scale into categories, such as high trust, somewhat trustful, and low trust, she could include the information in a footnote.

In the discussion of the sample you should identify the target population, sampling frame, sampling design, response rate, and when the data were collected. To avoid ambiguity, you should report the initial sample size, how many members of the sample were contacted, how many of those contacted belonged to the target population, how many refused to provide data, and how many supplied incomplete data. If possible you should compare respondents and nonrespondents. Any other sources of nonsampling error should be identified.

Findings

Whether you are writing a report or preparing an oral presentation the key considerations of how to present your findings are the same. You need to (1) organize the findings into a coherent presentation, (2) focus on the important findings and avoid overwhelming the audience with unnecessary detail, and (3) decide on how to present the data. Presented with an uninteresting analysis or an overwhelming amount of detail, audience members may stop listening or reading.

Your graphs and tables should complement the verbal presentation and exhibit data efficiently. Attractive graphics and clear explanations allow readers to assess the richness of the data. The location of graphics and explanations and the amount of space devoted to them signal the importance of the information they contain. You should not waste space on graphics that illustrate unimportant or trivial findings; they do not deserve major emphasis.

Tables may show exact numerical values. They are effective when you want to encourage many specific comparisons. Graphs are especially effective for time series and to make simple comparisons. They permit an audience: to pick out long-term trends, cycles, and seasonal fluctuations; to compare different groups or organizations; and to see differences before and after an intervention.

You should take care to avoid ambiguous labels. Spell words out and avoid abbreviations. The following summarizes practices associated with constructing effective tables and graphs.

1. Tables or graphs should have a precise, descriptive title. A title may list by name the dependent variable by independent variable by control variable (if any). Alternatively, a title may summarize a major finding supported by a graphic, for example, "City homicide rates have dropped over the past twenty years." All variables and their corresponding categories should be clearly labeled and appropriate units (e.g., years) should be indicated.
2. The independent variable normally heads the columns of a table and the dependent variable heads the rows.
3. If percents are used, the percent sign (%) should be entered at the top of each column.
4. The number of cases on which percent figures are based should be indicated. The total number of cases used in the analysis also should be indicated.
5. Statistical measures, if any, should be placed at the bottom of the table.

6. Definitions of key terms should appear as a table or graph footnote.
7. Data source(s) should be identified in the table or graph's footnote.
8. A good table supplements, not duplicates, the text. The table and its data should be referred to in the text, but you need to discuss only the highlights. As well, tables and graphs should be able to stand alone, that is, readers should be able to grasp the essential information without referring to the text.
9. As you work on preparing tables and graphs, remember to date them; you may even want to note the time. This is because as you analyze the data you may note and correct errors and you may decide on a different, more effective way to group your data. Unless your graphs or tables are dated you may not remember which represents the most recent version.

Discussion

In the findings section you report objective, verifiable information. In the findings section you organize and present the quantitative and qualitative data; in the discussion section you discuss what you observed about the information and interpret the findings. You may note

- what seems important;
- how the findings compare with the literature or stakeholders' perceptions;
- findings that were unexpected and your thoughts about why they occurred;
- implications of the findings for policy making, action, or further research.

Recommendations

Program evaluations, policy analyses, and other studies done for a legislative or administrative body may include recommendations. Recommendations are normative statements about changes that should be made in the program or policy. Although you may feel ill-equipped or uncomfortable in making normative statements or telling clients what they should do, the study's sponsors may expect recommendations.

You may find that recommendations focus decision makers on what needs to be done and increase the utilization of your findings. Recommendations should naturally follow from the research findings, that is, a reader should be able to figure out from the report's content why the recommendations were made. In making recommendations you should address only those changes the agency can make; for example, recommending a change in federal program requirements will not be of any value to a local social service agency. In some cases the costs and benefits of adopting a recommendation may be identified and included. Alternatively, you may suggest several options for agencies to consider.

Executive Summary

The executive summary highlights a report's content. The intended audience is the executive who has little time to read complete reports. An executive summary is also useful for many different audiences. Busy administrators and policy makers scan an executive summary to decide if and when to read the entire report or to refer it to an associate. Administrators with a limited interest in the topic skim a summary to keep themselves current. Policy actors may distribute summaries to communicate and endorse the report's findings. Investigators doing literature reviews can infer if the report is relevant and if they should read it.

Although usually included at the front of the research report, the executive summary is the last part of a report to be written. It includes only information contained in the report, but it can be read and understood independently of the report. In writing an executive summary, you decide what you want a reader to know. For example, you may visualize the impatient administrator who asks "What's the headline?" You can go through the report and find sentences that concisely describe why the study was done, who the subjects were, how the data were collected, major limitations in the methodology or its implementation, and what the major findings were. You should include any recommendations that were part of the report.

You should use clear direct sentences and visual cues to allow an individual to read the summary quickly. Keep its length and degree of detail consistent with the length and complexity of the report, agency expectations, and the importance of the findings. In preparing an executive summary, you should avoid including too many details; otherwise, its benefits are defeated. The following sample executive summary may serve as a model of how to organize and summarize a report.

ETHICAL ISSUES

Completing a research project, presenting findings, and storing information have ethical dimensions. A joint committee convened by the National Academy of Sciences, the National Academy of Engineering, and the Institute of Medicine identified three sets of ethical issues: research misconduct, questionable research practices, and other misconduct.[1] Research misconduct consists of acts of fabrication, falsification, or plagiarism. Questionable practices refer to decisions with regard to data retention and sharing, record quality, authorship, statistical analysis, and release of information. Other misconduct refers to acts that are unacceptable but not unique to researchers—for example, misuse of funds, vandalism, violations of government research regulations, and conflicts of interest. In this section we focus on research misconduct, handling research errors, and record-keeping issues as these are ones that you are most likely to encounter in the course of your work.

Research Misconduct

Fabrication is defined as making up data or results, and falsification as changing data or results. We assume that you know that fabrication is wrong. *Falsification* can be a bit more ambiguous. An easy way to falsify results is to drop cases from a dataset. Dropping selected cases can strengthen your statistical evidence or even

---

[1]Panel on Scientific Responsibility and the Conduct of Research, *Responsible Science: Ensuring the Integrity of the Research Process*, Report by the U.S. Committee on Science, Engineering, and Public Policy (Washington, D.C.: National Academy Press, 1992, Vols. 1 and 2).

AN EXECUTIVE SUMMARY

A Program Evaluation of the Vocational Training Programs at Portal

Portal [pseudonym] is a community-based rehabilitation facility whose mission is "to help people with vocational disabilities achieve a sense of self-worth by optimizing their potential to earn their own wages through work." The facility has two programs to provide persons with disabilities vocational training suited to their needs and abilities.

Study Questions

- How successful are Portal trainees in obtaining permanent jobs?
- What characteristics are associated with successful job placement?
- Is one of the training programs more successful than the other?

Findings

A customer satisfaction survey of trainees of both programs indicated that 22 of the 23 contacted were satisfied with the services they received; and were currently working in permanent positions.

An examination of Portal's databases found that based on case closure its success rate has declined in the past 3 years. In the first year it had 100 percent successes as compared to the state average of 87.2 percent. By the third year it had 81.3 percent successes as compared to the state average of 82.7 percent. The decrease may be attributed to a change in definition of the term *success*, which no longer considers probationary employment as a success. Analysis disclosed no differences in race, sex, or disability of clients in either program between those who were successful and those who were unsuccessful.

Inconsistencies in Portal's databases limited the evaluators' ability to find information and clarify definitions. A substantial amount of useful information was missing from the databases.

Recommendations

- Developing a centralized database for the entire agency with clear definitions of database fields, leaving little need for interpretation of information by data entry staff.
- Collecting more information on processes that could lead to better service for clients and employers.
- Connecting billing sheets to database to monitor hours of job development and job coaching for each client.
- Surveying or interviewing clients who have been placed in a permanent job to monitor their long-term success. ■

rescue a weak statistical model. For example, you might eliminate cases that you think have measurement errors, such as when you suspect that incorrect data were reported. If you think that measurement error occurred, you should try to confirm it. However, if you cannot confirm the error you may decide to remove the cases, in which case you must report that you removed the cases, explain why you removed the cases, and indicate how their elimination affected the results. The greater the effect, the more diligent you must be in reporting the decision. A decision that markedly affects the findings should not be buried in fine print.

You need not be overly concerned about avoiding charges of falsification. You simply need to be careful in documenting your decisions and why you made them. The documentation makes your decisions accessible for peer review.

Plagiarism is falsely presenting another's ideas or words as one's own. Quoted material should be placed in quotation marks and references cited. Closely following another author's diction is wrong. You should either use your own words and sentence structure or quote directly from your sources. Relying on the works of others is inevitable in research. No one knows this better than a textbook writer. We have referenced sources that we relied on to write segments of this text or that provided a unique or valuable perspective on the material. We have not referenced sources for ideas and perspectives that we know are part of the common knowledge of social science researchers.

Diligent referencing and use of your own words should be adequate to avoid charges of plagiarism. In our experience, the most common instances of plagiarism are using information from Web sites without citing sources, or editing and presenting another's work as one's own. Changing words and dropping sentences also constitutes plagiarism. If a report is to be published, you need to pay attention to copyright laws. You must get permission from the copyright holder to reproduce graphs, tables, long quotes, and other materials, including song lyrics, poetry, and cartoons. However, government documents are not covered by copyright, and their contents can be reproduced without obtaining permission. Nevertheless, you should use standard referencing procedures to cite a government document.

Handling Research Errors

Error is inevitable in research. The joint committee convened by the National Academy of Sciences, the National Academy of Engineering, and the Institute of Medicine identified four potential sources of error: the accuracy and precision of measurements, the generalizability of experiments, the quality of the experimental design, and the interpretation of the practical significance of the findings. To reduce the diffusion of incorrect knowledge, you should fully disclose your research procedures, and acknowledge and correct errors.

Full disclosure allows others to scrutinize the research. They may find errors by examining the research documents or by attempting to replicate the research. Concealing limitations in your methodology amounts to deception. Research reports should clearly identify and evaluate the limitations. In fact, the more troublesome a limitation, the more emphasis it should receive.

Complete information on research procedures can overwhelm readers with details and seriously diminish a report's effectiveness. The professional standards for program evaluation may serve as a useful guide. To provide useful information, the standards advise evaluators to write clearly, present information that their audiences can understand, and indicate the relative importance of their findings and recommendations. To achieve full disclosure, the standards advise evaluators to state their assumptions, their constraints, and how readers may obtain full information on research procedures, including data analysis.[2] The standards relieve evaluators of the burden of providing complete research information in every report, but they must take reasonable actions to ensure the accessibility of the database and documentation.

---

[2]The Joint Committee on Standards for Educational Evaluation, *The Program Evaluation Standards: How to Assess Evaluations of Educational Programs*, Second Edition (Thousand Oaks, CA: Sage Publications, 1994). Seventeen standards address reporting issues. The standards' application is not limited to educational studies.

SAVING DATA

Data must be saved and be accessible to allow research audits, replication of results, refinement of the analysis, additional analyses, or incorporation of data into other research designs Research data include completed data collection instruments, protocols for collecting and entering data, descriptions of experimental procedures, data files, computer printouts, field notes, videotapes, audio tapes, CDs, and DVDs. With this information the research can be reconstructed or replicated. Audits may be a component of ensuring integrity; auditors can investigate charges of falsification or fabrication.

Misunderstandings can be avoided if you and other involved parties agree on who will retain the data, how long they will be kept, and the conditions governing data sharing. Deciding who can access the data and issues affecting the confidentiality of respondents should also be agreed upon. If such agreements are not made explicit, administrators may find that they cannot access the data for further analysis. Some stakeholders, such as legislative bodies, may be unhappy if they cannot access and verify data collected by a funded program. Managers and researchers need to take into account others who may have legitimate reasons for reviewing the data.

Investigators may have discarded or misplaced data, information on how to retrieve data may not exist, or research documents may be scattered. If any of these situations exist, managers or investigators have not met their research obligations.

The exact amount of time for which data should be retained may depend on its assumed shelf life, agency customs, research agreements, or contracts. Completed questionnaires and records, needed primarily for research audits or to correct data-entry errors, may be kept the shortest time. Data files, data dictionaries, data collection instruments, and other research protocols may be saved indefinitely. Materials relating to a published work may be kept longer than those relating to an unpublished or in-house study. If an agency centralizes data collected on its behalf, it should develop a mechanism to exert quality control so that it does not get overwhelmed with useless data.

CONCLUDING OBSERVATIONS

Before presenting your findings, identify your audience, and decide how to get its attention and how to motivate its members to follow through on the report's findings and recommendations. Perusal of agency reports or listening to oral presentations may suggest how to organize information effectively for an agency.

Both oral and written presentations tend to include the same material, although the amount of time or detail will vary according to the audience. The sections of a research report are an overview or executive summary, a statement of the study's purpose, background information, the description of the methodology and results, along with supporting analyses, discussion, and, as appropriate,

recommendations. Important information should be clearly distinguishable and receive major emphasis. Remember that written reports may be passed around. You should not assume that every reader will know about a study's background or design. Consequently, you need to include necessary elaboration and documentation in the final written report. The methodology and findings sections in particular should be sufficiently documented so that the careful reader can make her own judgment on the adequacy of the report.

Oral presentations involve a captive audience. Still, the presenters are not guaranteed its attention. They should practice a presentation so that it flows logically and listeners can easily follow it. Visual aids should be easily seen and interpreted. If visual aids require special equipment, the presenters should know how to operate the equipment with minimal effort.

Your research reports should conform to ethical practice. Plagiarism, fabrication, and falsification clearly violate research ethics. Although human error is inevitable, you should take reasonable steps to minimize errors in your work. You should verify that that the correct data were entered, any modifications to the data were done correctly, and that the computer outputs were read and interpreted correctly. The descriptions of the study's methodology should disclose problems encountered, design limitations, and other information needed to facilitate critical review.

Where possible, you should let peers review your work. If you later find errors you are expected to acknowledge them.

When initiating a study, investigators and administrators need to agree on who will retain the data, how long data will be kept, and any conditions governing data sharing. During the study you should write up descriptions of research processes and decisions and label all research documents. Keeping a research log, where you record what was done, when, and by whom, will allow you to track and remember the various decisions and events that occurred. As the study ends, you should organize the data and documents and save them so others can use them to verify the research or conduct further analysis.

RECOMMENDED RESOURCES

Bardach, Eugene, *A Practical Guide for Policy Analysis: The Eightfold Path to More Effective Problem Solving*, Second Edition (Washington, D.C.: CQ Press, 2005). See especially "Tell Your Story," pp. 53–60.

Grob, George, "Writing for Impact," in *Handbook of Practical Program Evaluation*, edited by J. S. Wholey, H. P. Hatry, and K. E. Newcomer (2nd ed.; San Francisco: 2004, pp. 604–627) Grob's chapter covers the organization, presentation, and content of evaluation reports. The "Handbook" includes chapters on all aspects of program evaluation including research design, data collection, analysis, reporting and other topics.

Harrington, R., and Rakdal, S., *How to Wow with PowerPoint* (Berkeley, CA: Peachpit Press, 2007).

Publication Manual of the American Psychological Association, Sixth Edition, second printing, is the guide often used in the social sciences to cite sources. It offers guidance on the format for in-text citations, endnotes, references, reporting, and other topics.

Smith, Catherine, *Writing Public Policy: A Practical Guide to Communicating in the Policy Process*, Second Edition (New York: Oxford University Press, 2010). The book contains sections on the research process and sections on presenting information in a variety of written and oral formats.

Strunk, William, Jr., and White, E. B., *The Elements of Style*, Fourth Edition (New York: Longman, 2000) is a classic with rules for grammar and composition that have held up well for over 90 years. The book is available online.

The Chicago Manual of Style, Fifteenth Edition (Chicago: University of Chicago Press, 2003) covers referencing and copyright laws.

Walker, T. J., *How to Create More Effective PowerPoint Presentations* (New York: Media Training Worldwide, 2003).

A useful online resource for oral presentations may be found at http://www.the-eggman.com/writings/keystep1.html

CHAPTER 14 EXERCISES

This chapter has two exercises. In each you are asked to review and assess a presentation.

- Exercise 14.1 Learning from Practice: From Written Report to Oral Presentation asks you to read a written report, identify how it organized and presented the information, and outline an oral presentation based on the report.
- Exercise 14.2 Learning from Practice: Critiquing an Oral Presentation asks you to view and critique an oral report found on YouTube or another Web site.

EXERCISE 14.1   A Learning from Practice: From Written Report to Oral Presentation

Policy groups typically conduct research and present their findings in a variety of formats. To begin this exercise go to the Web site of such a group and download a research report. It can be reporting on public opinion, on program performance, or an evaluation. You may want to start your search at one of the following two sites.

The Urban Institute (www.urban.org)

The World Bank (http://econ.worldbank.org)

Section A: Getting Started

1. The Web site I visited was that of [name of organization and URL].
2. The name of the report I accessed was [name of the report].
3. The major divisions, section, or chapters of the report were [name(s) of divisions or chapters].
4. If it had an executive summary note how long it was. List in order its major sections.
5. What methodological information was included in the body of the report and in the appendices?
6. What did you notice about the tables and graphs contained in the body of the report?
   a. What types of graphs were used?
   b. How were the tables and graphs titled? (Did the report just name the variables or present a statement highlighting the findings?)
   c. What statistics were reported?
   d. Note the details included in a table or graph and comment on how the table or graph was discussed in the text. (Be sure to report the number and title of this graph or table.)
7. Were other research products produced from the same research project, for example, press releases? For each product found, summarize what parts of the research report were highlighted?
8. Identify a target audience that would be interested in listening to a presentation on the report. Justify your choice.
   a. Prepare an outline for a presentation.
   b. Create one or more PowerPoint slides to describe the study's methodology.
   c. Develop a list of tables and figures that you would include. Give each table or figure a descriptive title and briefly describe its format, for example, a bar chart.
   d. Create one or more PowerPoint slides of the findings that you would present.

Section B: Group Exercises

Group members who have selected similar reports (evaluations, citizen surveys, or academic papers) should work together in groups of two to four. Each group should develop a "Guide for Effective Written Reports" and a "Guide for Effective Oral Presentations."

1. Introduce each other to each report's purpose and intended audiences. From this information you may modify the guide's title to insert the type of report or audience, for example, "Guide for Writing Effective Reports of Citizen Surveys."

2. Compare each of the following components: organization of the executive summary and the report's organization and content (report sections; their order; use of visuals amount of detail in the discussion of methodology, tables, and figures; content and presentation of findings and recommendations; content of appendices). From this information develop your writing guide.
3. Compare and discuss each other's oral presentation outline: its organization, presentation of methodology, and presentation of findings. Based on this review develop your oral presentation guide.
4. Come together as a class and identify
   a. common elements in the guides;
   b. elements that are unique to one or more guides. Discuss whether these unique elements belong in guides for specific types of reports or if they should be included in all the guides;
   c. elements that seem to give contradictory advice. Discuss and try to resolve the differences: are they due to different types of reports, different target audiences, or different perceptions of what is effective?

EXERCISE 14.2   Learning from Practice: Critiquing an Oral Presentation

Go to YouTube. Find and watch a research presentation.
Perform the following assignments:

1. Describe and critique the presentation's organization.
2. Critique the presentation's clarity in establishing its purpose, describing its methodology, presenting its findings, discussing its findings.
3. Critique the presentation's use of visuals.
4. What advice would you give to the presenter(s) if the presentation was to a professional group instead of to a general audience? Would your advice be different if the presentation is to a legislative body?

GLOSSARY

Accuracy   How close the sample statistic is to the value of the population parameter it is used to estimate.

Anonymity   Collecting information so that researchers cannot link any piece of data to a specific, named individual.

Arithmetic mean   A measure of central tendency determined by adding together the values of a variable for all cases in the distribution and dividing this sum by the total number of cases. The use of the arithmetic mean requires that data be measured at the interval or ratio level.

Attribute data   Data other than geographic data about a place used in GIS.

Biased question   A survey question that elicits inaccurate information, because it is worded in such a way that respondents are encouraged to give one answer rather than another.

Classical experimental design   A design with at least one experimental group and one control group, with subjects randomly assigned to each group by the investigator, the independent variable under the control of the experimenter, and a pretest and posttest.

Closed-ended questions   A type of survey question in which the respondent is given a list of possible answers and is requested to select an answer or answers from that list.

Community forum   A meeting in which community members and stakeholders have the opportunity to discuss issues facing the community and exchange data from a cross-section of the community in a neutral setting.

Comparison group design   A quasi-experimental design similar to the classical experimental design. However, the researcher does not randomly assign subjects to groups and the occurrence of the independent variable may not be controlled by the researcher.

Confidence interval   Interval around a statistical value within which the parameter is likely to be found. A confidence interval of 95 percent is most common.

Confidence level   The confidence that an investigator has that a sample estimate is within a specified range of the parameter. A confidence level of 95 percent is most common.

Confidentiality   Protection of information, so that researchers cannot or will not disclose records with individual identifiers.

Contingency table   A table showing the relationship between two or more variables. Usually used for nominal and ordinal variables. The table shows how the distribution of one variable is related to the values of another variable.

Control variable   A variable included in an analysis to determine if it affects the relationship between two other variables. The values of the control variable are held constant while the relationship between the other two variables is evaluated.

Correlation coefficient   A statistic that measures the strength of association between two or more variables. If the variables are ordinal, interval or ratio, the statistic also indicates direction. The most common correlation coefficient, Pearson's r, is used with interval and ratio variables and measures strength and direction.

Cost benefit analysis   A type of analysis that compares the dollar value of the costs of a one or more programs or courses of action to the dollar benefits of those programs or courses of action.

Cost effectiveness analysis   A type of analysis that compares the costs and outcomes of two or more courses of action. Costs of both courses of action are expressed in dollar values and outcomes are compared as to magnitude. Unlike cost-benefit analysis outcomes are not valued in monetary terms.

Cross-sectional design   A type of research design with all relevant variables measured at the same time.

Decision map   A map used by program analysts and managers to determine which aspects of a program are successful in particular areas.

Deductive disclosure   Disclosure that occurs when information from a research project can be analyzed to identify a specific person or otherwise compromise the privacy of the subjects.

Density mapping   Process of mapping for an area the number of instances of an event or cases with a given value relative to other areas. The area rather than the incident or case is the unit of analysis.

Dependent variable   A variable whose values a researcher wants to explain; an outcome or result of an independent variable.

Descriptive statistics   Statistics that describe and summarize data without making inferences to a larger set of cases.

Efficiency   Ratio of the amount of input to the amount of output.

Equity   A value that services should be available to all eligible persons regardless of race, ethnic group, religion, gender, or sexual orientation.

Evidence of operational validity   Evidence that a researcher is measuring what he or she wants to measure. Common evidence is based on (1) a review of an operational definition's content to ascertain that it is relevant and representative and (2) empirical comparison of a measure with a similar alternative measure or with a predicted outcome.

Executive summary   Written summary highlighting the main ideas of a report; it should contain only information from the report, but be understandable independent of the report.

Experimental design   A type of design in which the researcher can assign subjects to different research groups, control who is exposed to the independent variable, when they are exposed to it, and the conditions under which the experiment takes place.

Expressed need   A need that is given voice as a demand or request.

External validity   The extent to which a study produces findings that apply to cases not in the study.

Fabrication   Making up data or results.

Falsification   Changing data or results. This can include dropping certain cases in a dataset to strengthen statistical evidence.

Filter question   A type of survey question used to identify respondents who should answer specific follow-up questions.

Focus group   A research tool utilizing small-group interviews to obtain qualitative data as well as items for questionnaires or surveys. Group interaction is an important component of a focus group.

Forced-choice question   A closed-ended question that requires the respondent to choose among available options with no provisions for an "other" or "none of the above" response.

Frequency distribution   A table or graph listing the variable values along with the number of cases with each value.

Generalizing map data   Smoothing or leaving out certain levels of mapping details.

Geocoding   The assignment of a code—usually numeric—to a geographic location. Geocoding provides GIS software the ability to find an address or other coded location in a database and place a symbol, usually a dot, on a map to indicate that location.

GIS (Geographic Information Systems)   Computer systems used for inputting, storing, manipulating, analyzing, and displaying geographic data.

Histogram   A type of bar graph in which the widths of the bars represent the range of values included in a category and the heights of the bars represent the number or percent of cases with those values. Similar to a bar chart except that both the length and width of the histogram bars have meaning. Histograms can be used for interval and ratio variables.

Hot spot mapping   Using smoothed points of activity to create density areas in GIS that represent the probability of an event's occurrence.

Hypothesis   An empirically testable statement describing the relationship between two variables.

Independent variable   A variable that is assumed to explain or cause variations in another variable, the dependent variable.

Inferential statistics   Statistics that estimate the values of population parameters from a probability sample.

Informed consent   The principal that (1) prospective subjects of research must be informed of the purpose of the research and of any risks that may be incurred by participating, and (2) the subjects must give their explicit consent prior to participating in a study. Subjects may be told of benefits of participating.

Institutional Review Board (IRB)   An internal body created by institutions that receive federal money for research involving human subjects; the IRB reviews all institutional research involving human subjects to determine whether it conforms to ethical practices.

Interagency map   A map containing combined information from multiple agencies.

Internal consistency   Evidence of reliability that establishes if the items in a measure with multiple items are empirically related to each other.

Internal validity The extent to which a research design and its implementation provide evidence that a specific independent variable caused a change in a dependent variable.

Inter-quartile range The range of values encompassing the middle one-half of the observations in an ordered distribution.

Interrupted time-series design A quasi-experimental design incorporating an independent variable other than time into the time-series design. Several values of the dependent variable are recorded before and after the occurrence of the independent variable.

Linear relationship A relationship between two variables that can be described by a straight line. As the value of one variable increases, the value of the other variable increases or decreases at a constant rate.

Loaded question A biased question worded in such a way that the respondent perceives that only one way of answering is acceptable.

Logic model A model in which inputs, organizational activities, outputs, immediate outcomes, and a long-term outcome are diagrammed to illustrate how they relate to each other and are expected to determine a final outcome.

Measure of dispersion Measures that indicate the extent to which values in a distribution differ from each other. These include the range, inter-quartile range, percentiles, average deviation, variance, and standard deviation.

Measurement Assigning a value to variables for each case.

Measures of central tendency Measures indicating the value of a distribution that is representative, most typical, or central. These measures include the mode, median, and arithmetic mean.

Median A measure of central tendency; the value of the middle case of an ordered distribution of values. The median requires ordinal, interval or ratio variables.

Meta-analysis A systematic technique utilized by researchers to analyze a set of existing studies. It is conducted to draw general conclusions from several empirical studies and to identify hypotheses that merit further testing.

Mode A measure of central tendency; the value of a variable that occurs most frequently.

Multiple regression Regression in which two or more independent variables are included in a regression equation and analyzed to assess their relationship to the dependent variable.

Nominal scales Scales that categorize and label the values of a variable. The investigator can group cases by the variable categories but cannot order them.

Non-experimental design Designs that lack most if not all of the characteristics of a design that would provide evidence of causation. These designs do not control for the threats to internal validity.

Nonprobability sampling Type of sampling in which case selection is based on criteria other than random selection. Hence, the probability that any case or set of cases will be in the sample is unknown. Availability, purposive, quota, and snowball are commonly is used nonprobability sampling designs.

Open-ended question A type of survey question in which respondents are required to provide their own answers without a listing of possibilities from the researcher.

Operational definition The specific set of rules and procedures used to assign values to variables.

Operational validity Judgment that a measure is named correctly and is measuring what it is said to measure.

Ordinal scales A measurement scale that orders the values of a variable and allows an investigator to order the cases based on their variable value. Ordinal scales do not, however, measure the quantitative difference between cases.

Parameter Characteristic of the units in the population. A primary benefit of probability samples is the ability to estimate parameters from sample statistics.

Pattern detection The process of looking for similar events, circumstances, classes of individuals, and so on within a geographic area.

Perceived need Something people consciously lack and desire.

Percentage change A relative frequency that converts to a percentage the amount of change in the value of a variable from one time to another for the same case.

Percentage difference The difference between the percentages of cases in different categories of the independent variable for a value of the dependent variable.

Periodicity The regular occurrence of an event or characteristic. Periodicity is a possible problem in systematic sampling if events occur on the same schedule as the skip interval

Pilot study A small study designed to test the adequacy of a proposed data collection strategy.

A pilot study should test planned measures and analysis on a sample that represents the target population. Pilot study may also refer to demonstration projects to implement an innovative program or treatment.

Plagiarism Falsely presenting another's ideas or words as one's own.

Population Total set of units you are interested in studying; a sample is selected from this set of units.

Presentation maps Maps that include a persuasive point or message the author wishes to highlight.

Pretest An initial test of a proposed questionnaire or other survey instrument given to a small group of subjects who represent common variations found in the target population.

Probability sampling Type of sampling in which each unit of the population has a known, nonzero chance of being in the sample. The following are the basic probability sampling designs: simple random; systematic; stratified; cluster; multistage.

Productivity Ratio of the amount of output to the amount of input.

Proximity map A map illustrating how far something is in relation to something else.

Qualitative evidence of reliability Evidence based on a review of directions, a measure's wording and response choices. Also, involves determining the availability of the information to respondents.

Qualitative research Research involving detailed, verbal descriptions of characteristics, cases, and settings. Qualitative research usually involves relatively few cases investigated in depth.

Quantitative evidence of reliability Statistical measures that commonly include inter-rater, test-retest, and equivalent forms reliability. Quantitative evidence consists of the correlation between different forms of a measure, different researchers using the same measure, and of a set of items all purporting to assign a value to a variable.

Quantitative research Research in which values of variables are characterized by numbers. Data are summarized and analyzed with statistical techniques. Typically research is done on many cases and includes many variables.

Quasi-experimental design A design having some but not all of the characteristics of the experimental design. Often used in applied-research situations when random assignment and the control necessary for a true experiment is not possible or practical.

r^2 **(r squared)** The square of the Pearson's correlation coefficient measuring the percentage of variation in one variable that is statistically related to variation in another variable.

Randomized posttest design An experimental design with at least two groups and a posttest but no pretest. Subjects are randomly assigned to groups.

Range A measure of dispersion giving the quantitative distance between the lowest and the highest values in an ordered distribution.

Rate A relative frequency used to standardize the occurrence of some event. Rates are calculated by dividing the frequency of occurrence of an event by a total frequency count, such as the population of a jurisdiction in which the event takes place. The result is multiplied by a base number that is a power of 10, such as, 10, 100, or 1,000.

Ratio scales A measurement scale that allows researchers to rank subjects and determine the exact quantitative difference between them. Ratio scales have an absolute zero point. In administrative work almost all numerical scales are ratio.

Regression coefficient The b coefficient in a regression equation. It tells us the amount and direction of change occurring in Y, the dependent variable, for a unit increase in X the independent variable.

Regression equation An equation for a straight line that best estimates the relationship between two ratio variables. The form of the equation is $Y = a + bX$.

Reliability Evidence that a measure is relatively free of random error. Reliability also refers to the consistency of a measure and allows one to assume that differences between subjects or over time are real differences.

Response rate Percentage of individuals or organizations selected for a sample that answer a survey or agree to be interviewed.

Sampling frame The list of members of the population from which the sample is taken.

Secondary data Existing data that were collected for a purpose other than for the given study.

Threats to external validity Factors that limit the ability to apply the findings of a study to cases not involved in the study.

Threats to internal validity Factors in a study other than the independent variable that could be the cause of a change in a dependent variable and represent alternatives to the tested independent variable.

Sample   A subset of units selected from a larger set, the population. Information from the subset is used to find out something about the population.

Sampling error   Mathematical estimate of the difference between a parameter value and a statistic used to estimate it.

Sampling unit   Unit or set of units selected for a sample.

Scatterplot   A figure showing the relationship between two ratio variables. With the independent variable values on the horizontal axis and the dependent variable values on the vertical axis, the values of x and y for each case are plotted. The resulting plot graphically represents the relationship between the variables.

Sensitivity   The characteristic of a measure indicating how well it distinguishes between individual cases and small differences in values. A sensitive measure identifies small differences; a less sensitive measure does not.

Statistic   Summary values of the characteristic of units in a sample. We use sample statistics to estimate values of population parameters.

Time series   A set of data obtained at regular intervals for a single case; a type of line graph with the units of time displayed along the horizontal axis and the values of a variable or frequency of an occurrence on the vertical axis.

Unit of analysis   Case or subject whose characteristics are measured and examined. This is the smallest level at which data are collected.

Variable   A characteristic that has more than one value.

INDEX